P9-ECM-134

International Brand Packaging Awards 2

ROCKPORT PUBLISHERS • ROCKPORT, MASSACHUSETTS

First published in the United States of America by:
Rockport Publishers, Inc.
146 Granite Street
Rockport, Massachusetts 01966
Telephone: (508) 546-9590
Fax: (508) 546-7141

Distributed to the book trade and art trade in the U.S. and Canada by:
North Light, an imprint of
F & W Publications
1507 Dana Avenue
Cincinnati, Ohio 45207
Telephone: (513) 531-2222

Other Distribution by:
Rockport Publishers, Inc.
Rockport, Massachusetts 01966

ISBN 1-56496-154-0

10 9 8 7 6 5 4 3 2 1

Art Director: Laura Herrmann
Designer: C.A. Buchanan Design

Printed in Hong Kong

Contents

Preface

The reason that Graphic Design:USA has singled out package design and brand identity for a sponsored competition—from among all of the design disciplines—is threefold.

First, package designers are "ahead of the curve" in marshaling design in support of marketing strategies and strategic brand objectives. They can show the way to other designers who must master this skill to fully make their power felt in the broader business community. Second, package designers uniquely face two critical challenges that all communicators must grapple with in the '90s: piercing through the clutter of an information-sated age and creating a message that is relevant in an increasingly global marketplace. This competition could help clarify these challenges and suggest creative solutions. Third, package design plays a vital role in an economy where point-of-purchase is more and more critical—informing, educating, selling, directing, decorating—and its practitioners deserve recognition and encouragement.

Based on the quantity and quality of submissions to our second competition—and the extraordinary array of winners displayed in this book—the International Brand Packaging Awards has exceeded even our most brazen expectations. The best work reproduced here affirms the marriage between design and marketing more intently than ever. Also, solutions to problems raised by clutter and by the globalization of marketing are intelligent and effective. This book, as well as a special roundup of winning pieces in our monthly magazine, provides a meaningful showcase for first-rate work of a deserving group of professionals.

In Graphic Design:USA's January 1994 issue, the Boston graphic designer Robert Cipriani looked at the marketing revolution now underway, and concluded that "[d]esigners still have the responsibility to create communications between human beings that solve problems, disseminate information, and transmit emotion, grace, style and order." This book is eloquent testimony that a select few people are more than meeting that responsibility.

Gordon D. Kaye

Editor and Publisher,

Graphic Design:USA Magazine

Introduction

The gratifying response to this Second Annual International Brand Packaging Awards demonstrates that package designers are thinking globally and responding with insight and understanding to the marketing challenges of the 1990s. The diversity of the entries—from the U.S., Canada, Latin America, Europe and Asia—further enhances the stature of International Brand Packaging Awards as the premier global packaging and brand identity design competition.

The purpose of this worldwide program is to provide a forum for package design professionals and to build recognition for their contributions to the marketing process. These contributions take on more meaning when the designer is brought into the situation during the early planning stage for a new product launch or the repositioning of an existing brand. By coordinating the skills of the designer with those of the advertising agency and the client, as a team, the package can work as a powerful stimulus for motivating a purchase decision by the consumer. When the package design is treated as an afterthought, the results generally are less successful.

From among the thousands of entries, our distinguished panel of judges selected 180 Best of Show, Gold, Silver, Bronze, and Honorable Mention Awards. The hundreds of packages and logotypes that you will become acquainted with on the following pages, including many which were chosen as worthy of special consideration although they were not winners, prove that good design can be compatible with sound marketing principles.

We extend our thanks and best wishes to the hundreds of designers throughout the world with whom we have formed close relationships and whose participation is crucial to this unique competition. And, special thanks to Parsons School of Design in New York for honoring International Brand Packaging Awards with a month-long exhibition, which served as a learning experience for its students as well as a celebration for the design community.

Murray J. Lubliner

Director

International Brand Packaging Awards

The Awards

International Brand Packaging Awards is recognized as the foremost global competition that focuses on how independent and company designers contribute to the solution of marketing problems through packaging and brand identity design.

Designers, advertising agencies, and marketers may submit packages, labels, lines of packages, and brand logotypes produced in the previous year. This program presents a unique opportunity for designers to share their best work with the global business community and with colleagues around the world.

The IBPA event was begun and is directed by Murray J. Lubliner, who has played a pivotal role in the development of packaging and brand identity strategies for marketers in the U.S. and other countries. Among his areas of expertise are multinational marketing, corporate and brand identity, and naming products and companies.

Lubliner considers packaging the critical and final link between brand image advertising and the consumer's purchase decision at the point-of-sale. When packaging is integrated with advertising strategy and in-store merchandising, the chances for the brand's success is increased significantly.

Among the companies that have taken advantage of Lubliner's experience are Allied-Signal, American International Group, ConAgra, Gerber, Hershey, Interpublic Group of Advertising Agencies, Shell Oil, Transamerica, and Warner- Lambert.

Before forming Murray J. Lubliner Associates, based in New York, Lubliner served as a founding partner of Lubliner/Saltz Inc. for 17 years and as a senior marketing executive of Lippincott & Margulues. He is a graduate of New York University, where he is an adjunct professor at the NYU Management Institute and teaches "Image, Identity and the Bottom Line." He speaks and writes extensively on brand and corporate identity issues.

The Judges

The Second Annual International Brand Packaging Awards is proud to highlight the distinguished panel of design, communications, and research professionals who evaluated each entry on how effectively the package, label, or brand identity supports brand marketing strategy. Each participant in IBPA is asked to write a brief description of marketing objectives along with each entry. The judges are

Terry Schwarts
Director of Packaging Communications
Kraft General Foods, Inc.

Elizabeth A. Morales
Art Production Manager
Nabisco Foods Co.
Package Design Dept.

Owen W. Coleman
President
Coleman, Lipuma, Segal & Morrill, Inc.

Ralph Colonna
Principal
Colonna Farrell: Design Associates

Michael Penrod
Director of Corporate Design
Colegate-Palmolive Co.

Howard J. Alport
Principal
Lipson-Alport-Glass & Associates

Herbert Meyers
Managing Partner
Gerstman + Meyers Inc.

Linda M. Keefe
Manager of
Corporate Identity
3M

Elliot Young
President
Perception Research Services, Inc.

Terry Schwarts
Director of Packaging Communications
Kraft General Foods, Inc.

Elizabeth A. Morales
Art Production Manager
Nabisco Foods Co. Package Design Dept.

Owen W. Coleman
President
Coleman, Lipuma, Segal & Morrill, Inc.

Ralph Colonna
Principal
Colonna Farrell: Design Associates

Michael Penrod
Director of Corporate Design
Colegate-Palmolive Co.

Howard J. Alport
Principal
Lipson-Alport-Glass & Associates

Herbert Meyers
Managing Partner
Gerstman + Meyers Inc.

Linda M. Keefe
Manager of Corporate Identity
3M

Elliot Young
President
Perception Research Services, Inc.

Package Land Co., Ltd.
Osaka Japan
Yasuo Tanaka
Clover Co., Ltd.

Coley Porter Bell
London UK
Colin Porter
Alison Cane
The Great Atlantic & Pacific Tea Co., Inc.

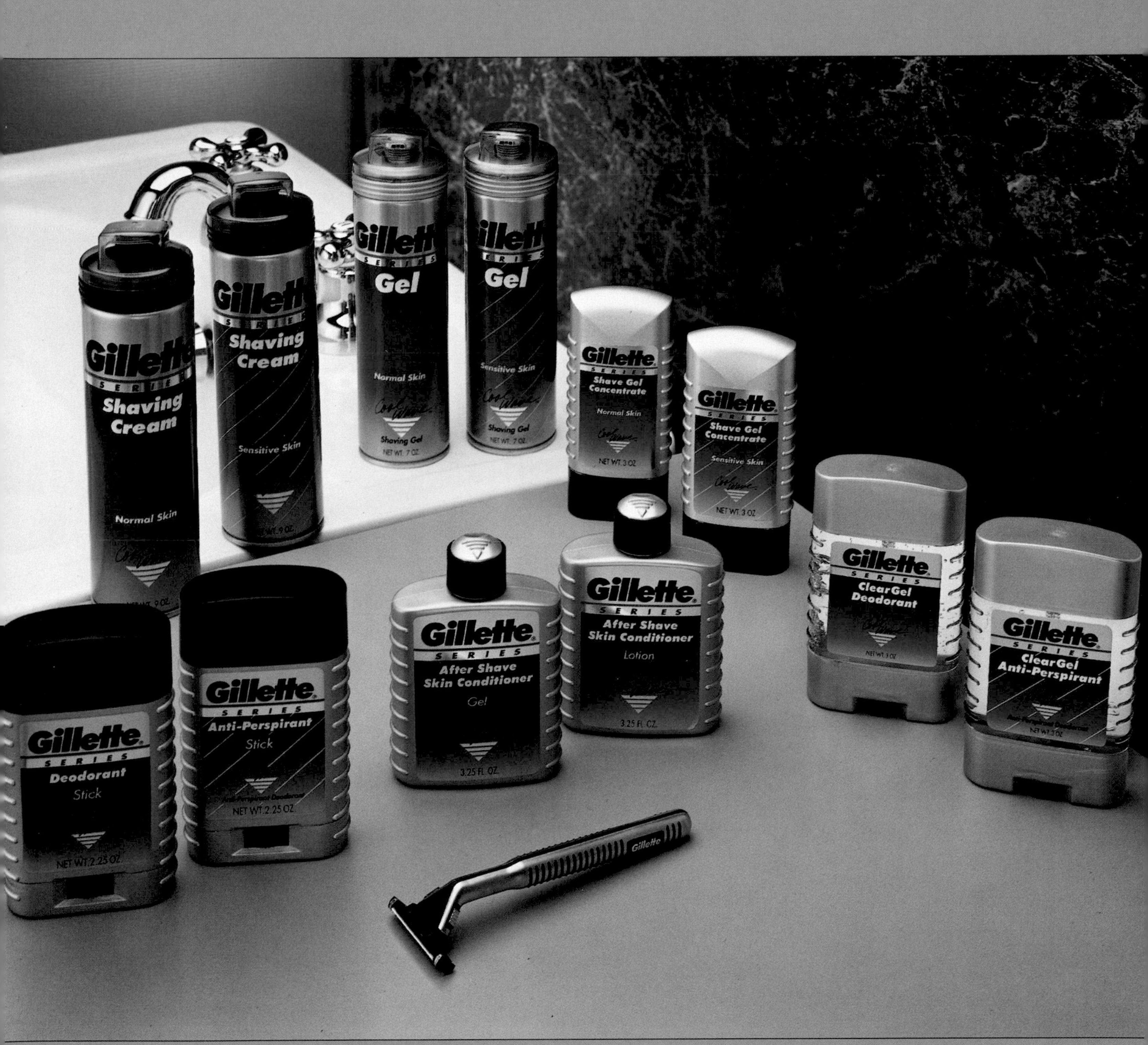

Desgrippes Cato Gobe & Associates

New York NY

Marc Gobe

Tim Robinson

Gillette

The BoardRoom Design Group
Cleveland OH
Kim Zarney
Bob Wood
Peter Renerts
Anchor Hocking Glass Company

Earl Gee Design
San Francisco CA
Earl Gee
Fani Chung
John Mattos
Quorum Software Systems, Inc.

Lewis Moberly
London UK
Mary Lewis
Amanda Lawrence
John Sugden
ASDA Stores

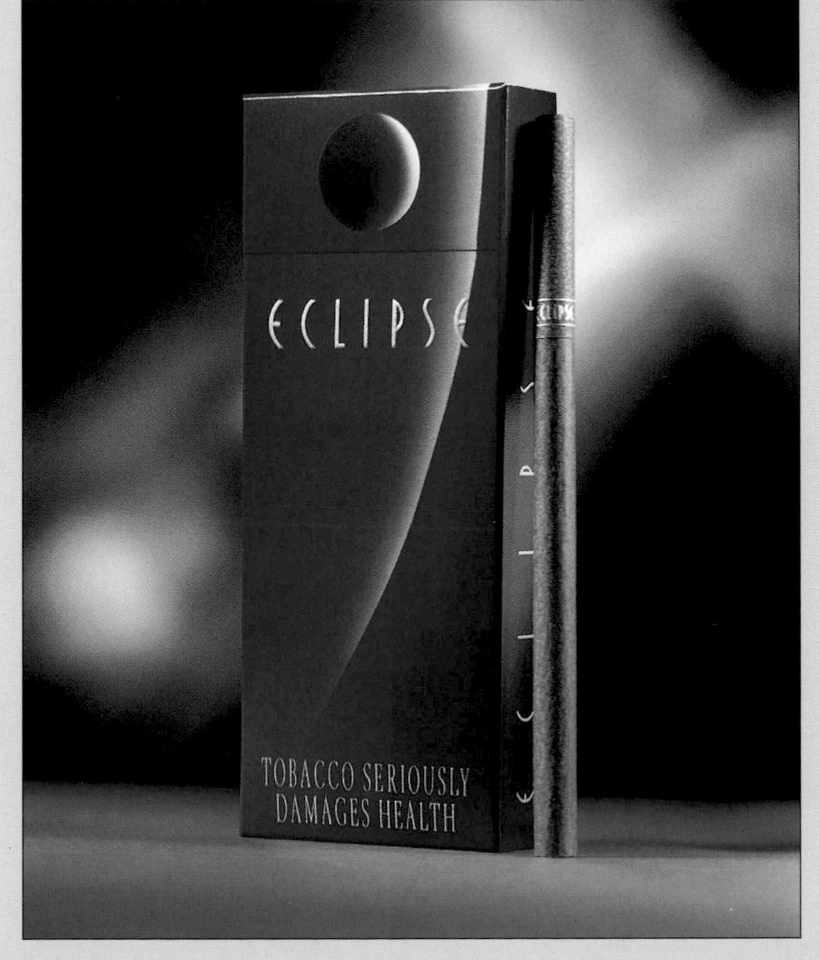

Light & Coley Limited
London UK
Laurie Light
Martin Lewsley
Nigel Tudman
Gallaher Limited

HORNALL ANDERSON DESIGN WORKS, INC.
Seattle WA
JACK ANDERSON
DAVID BATES
LIAN NG
Giro Sport Design Inc.

WHITE TIGER, INC.
Audubon NJ
THYRA O'BRIEN
SCOTT BOWKER
Thomson Consumer Electronics, Inc.

MILLEN & RANSON INC.
New York NY
MORTON MILLEN
PAUL RANSON
ROBERT JOHNSON
New York Philomusica

Design Board Bahaeghel & Partners

Brussels Belgium

Christophe Blin

Thierry Borremans

Denis Keller

Elf Aquitaine

Davies Hall
London UK
Tim Leslie-Smith
Robin Hall
PetWorld

Agnew Moyer Smith, Inc.
Pittsburgh PA
Grant W. Smith
Rany Ziegler
Ingrid Nagin
The Iams Company

Lewis Moberly
London UK
Mary Lewis
Bruce Duckworth
Stefano Fabrucci
La Rinascente

Harte Yamashita & Forest
Los Angeles CA
Susan Healy
Knott's Berry Farm

Midnight Oil Studios
Boston MA
Kathryn A. Klein
James M. Skiles
James T. McGrath
Genfoot

Lewis Moberly

London UK

Mary Lewis

Jimmy Yang

Lucilla Scrimgeour

Karl Henecka

Bahlsen Keksfabrik GmbH

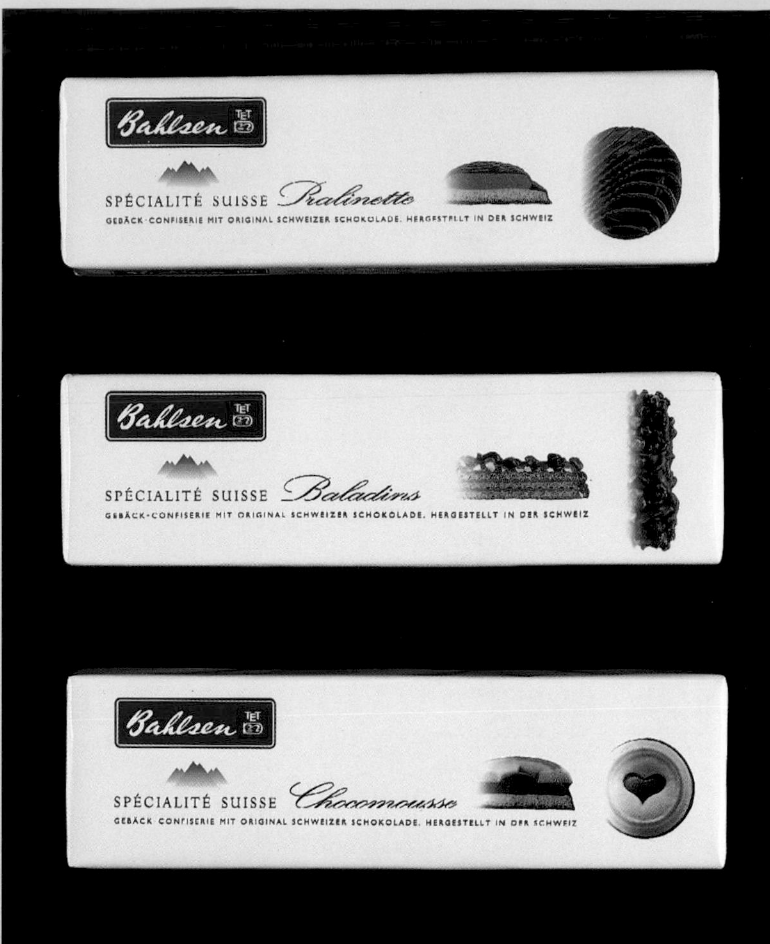

Design Bridge (UK) Ltd.

London UK

Paul Browton

Wendy Mellors

Laurie Evans

Crown Berger Ltd.

Design Board Bahaeghel & Partners

Brussels Belgium

Sally Swart

Denis Keller

Martin Van Grevenstein

S.C. Johnson

BaileySpiker, Inc.

Norristown PA

Christopher K. Bailey

Paul D. Spiker

Dana C. Winslow

FreshWorld Farms

Midnight Oil Studios

Boston MA

James M. Skiles

Kathryn A. Klein

James T. McGrath

Parker Brothers

Gold

United States Surgical Corporation
Norwalk CT
Anita Patterson
Kevin A. Williams
Timothy Ahrens
MArgaret Saleeby
United States Surgical Corporation

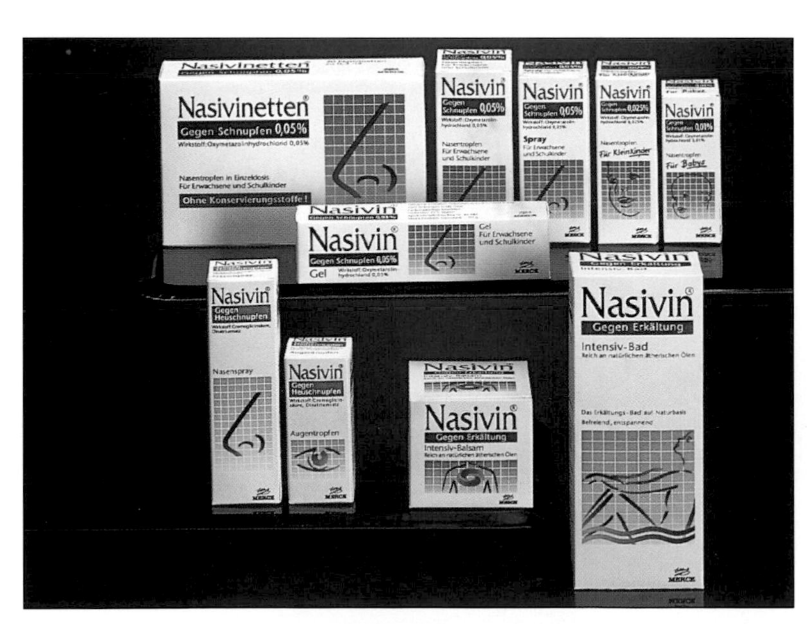

Hartmann & Mehler Designers GmbH
Frankfurt Germany
Angela Spindler
Bernd Wilhelm
Merck Produkte-Vertriebsgesellschaft & Co.

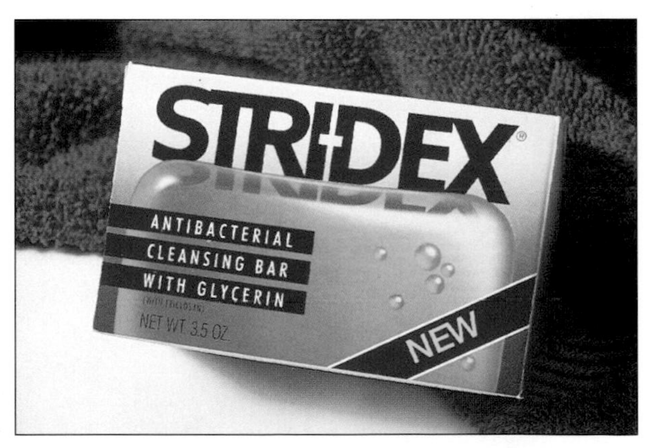

Hans Flink Design Inc.
White Plains NY
Hans D. Flink
Stephanie Simpson
Suzanne Clark
Jaque Auger
Sterling Health USA

Kose Corporation

Tokyo Japan

Fujio Hanawa

Tenji Motoda

Junichiro Kaneko

Kose Corporation

Fisher Ling & Bennion

Cheltenham UK

Jane Myhill

Debbie Sanders

Javier Sanchez

HP Bulmer

SOLARIS

The Design Company/San Francisco
San Francisco CA
Barry Deutsch
Richard Scheve
Sandra Koenig
The Solaris Group

Neumeier Design Team
Palo Alto CA
Marty Numeier
Christopher Chu
Curtis Wong
ZSoft Corporation

Monnens-Addis Design
Emeryville CA
Steven Addis
Joanne Hom
Hugh Howie
E&M Games

Schafer Associates, Inc.
Oak Brook Terrace IL
Barry Vinyard
Jill Ingrassia
Karen DiMonte
Tony Cesare
Suburbia, S.A. de C.V.

Design Partners
Toronto Canada
Frank Gottvald
Sandra LeBlanc
Platex Ltd.

PRIMO ANGELI INC.
San Francisco CA
PRIMO ANGELI
PHILIPPE BECKER
DERICK GRINELL
Just Desserts

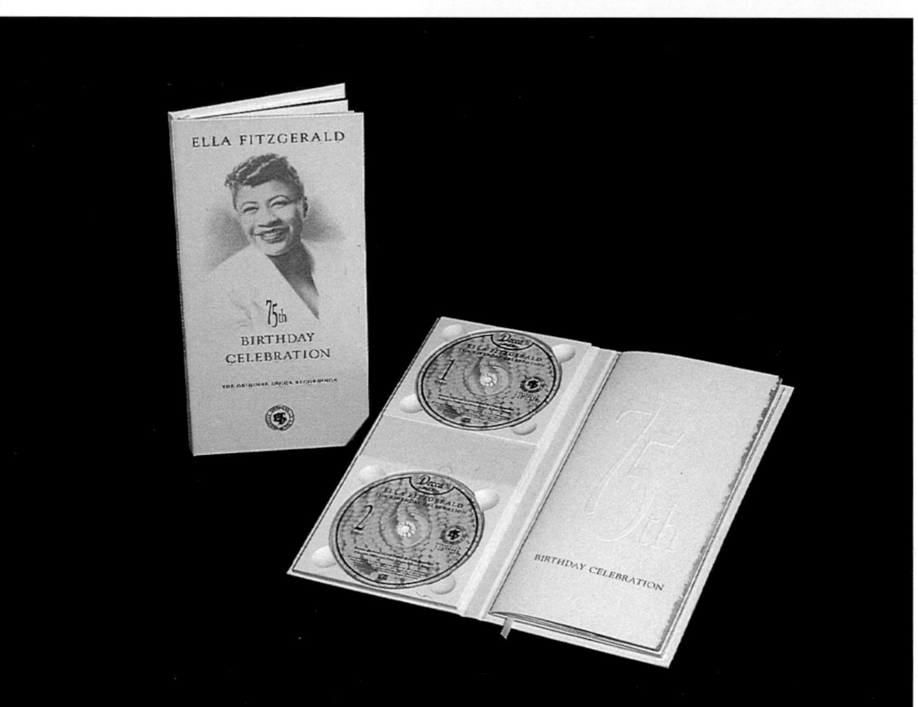

G.R.P. RECORDS
New York NY
ANDY BALTIMORE
SCOTT JOHNSON
BUD KATZEL
DON MORGENSTERN
Ella Fitzgerald/Decca

WALSH AND ASSOCIATES, INC.
Seattle WA
MIRIAM LISCO
KATIED DOLEJSI
GLENN YOSHIYAMA
Vitalli

Profile design/KDF Planet
San Francisco CA
Kiyoshi Komaki
Masayo Sakamoto
Kanetetsu Delica Foods

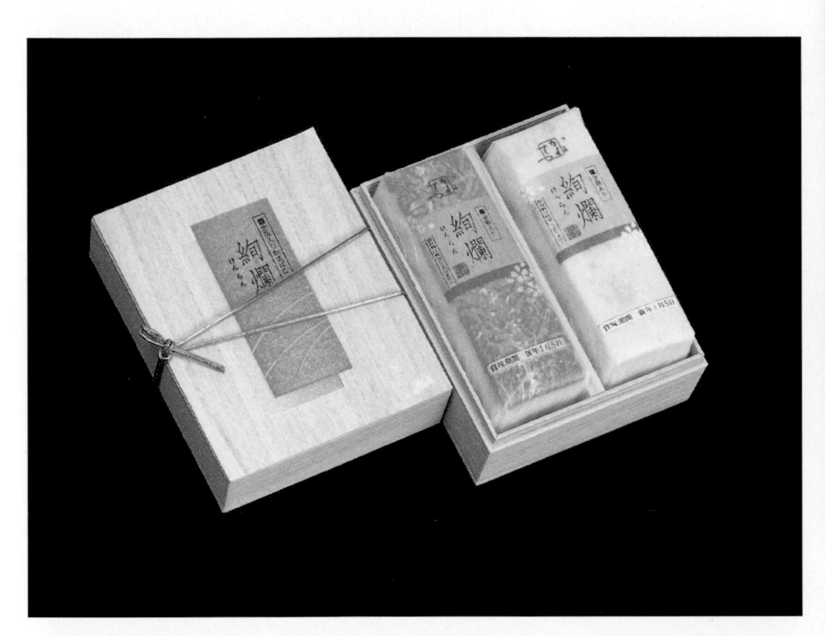

The Thompson Design Group
San Francisco CA
Dennis Thompson
Jody Thompson
Elizabeth Berta
Buena Vista Winery

Ostro Design
Hartford CT
Michael Ostro
Neil Shigley
Peter Dunn
Canson-Talens, Inc.

Pineapple Design S.A.

Brussels Belgium

Rowland Heming

Richard Evans

Fina Europe

Siebert Head

London UK

John Parsons

Tony Watts

Dean Gower

Texaco Services (Europe) Ltd.

R. Bird & Company Inc.

White Plains NY

Joseph Favata

Lever Brothers Co.

Coley Porter Bell

London UK

Simon John

Karen Dunbar

Pitman Moore

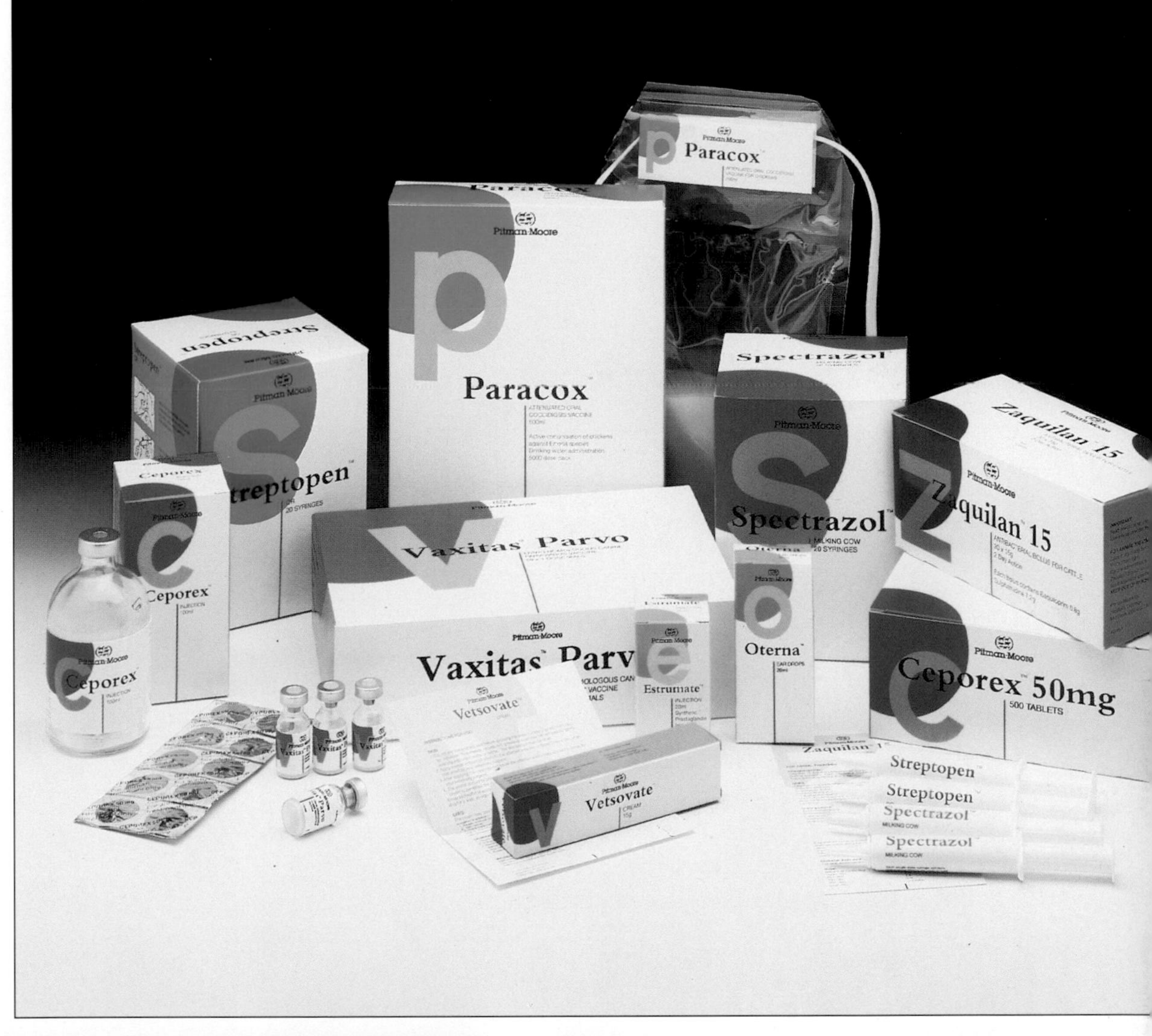

Miller Sutherland

London UK

Kathy Miller

David Harrison

Germaine De Capuccini

The Schechter Group

New York NY

The Schecter Identity Design Grp.

Gary Stilovich

Sterling Winthrop Inc.

MORILLAS & ASOCIADOS
Barcelona Spain
Royal Brands, S.A.

COLEMAN LIPUMA, SEGAL & MORRILL, INC.
New York NY
OWEN W. COLEMAN
RICHARD C. ROTH
CANDY CACIOLO
Ralston Purina Co.

RATTA DESIGN COMMUNICATIONS
Portland ME
DONNA HUNTER
ROBERT RATTA
BRUCE HUTCHISON
Cozy Harbor Seafood, Inc.

FISHER LING & BENNION
Cheltenham UK
JANE MYHILL
ANDREW HEATH
Selfridges

DESIGN BRIDGE (UK) LTD.
London UK
ROD PETRIE
DAVID ANNETTS
DANIEL CORNELL
Bacardi Eurpoe

THE SCHECHTER GROUP
New York NY
RONALD WONG
Liggett Group

Schafer Associates, Inc.
Oak Brook Terrace IL
Barry Vinyard
Jill Ingrassia
Karen DiMonte
David Koe
Bernhard Woodwork, Ltd.

Graphic Edge
Huntington Beach CA
Ryan Rieches
Susan Campbell
Washington Forge

J. Works Co., Ltd.
Osaka Japan
Jun Saeki
Minako Ueshima
Hills Bros. Coffee Japan

The Green House
London UK
Judi Green
Brian Green
Forte Restaurants

Charles Zunda Design Consultants Inc.
Greenwich CT
Charles Zunda
Sylvia Lorenz
Charles Zunda Design Consultants Inc.

David Scarlett Associates, Inc.
New York NY
David Scarlett
The American Tobacco Company

Bright & Associates
Venice CA
Keith Bright
Raymond Wood
Mark Verlander
Miller Brewing

Parham Santana Inc.
New York NY
Millie Hsi
Richard Tesoro
Beacon Looms, Inc.

The Great Atlantic & Pacific Tea Co., Inc.
Montvale NJ
Tony Lento
Karen Welman/Sterling Design
Michael Jackson/C.M. Jackson Co.
The Great Atlantic & Pacific Tea Co., Inc.

The Leonhardt Group
Seattle WA
Susan Cummings
Bruce Hale
Sea Pen Press
Petros Winery

Landor Associates

San Francisco CA

Quentin Murley

Jon Weden

Michael Livolis

Katherine Salintine

H.E. Butt Grocery Co.

Elmwood

Leeds UK

Clare Marsh

Julia White

Julian Hayes

Macaw (Soft Drinks) Ltd.

Hunt Weber Clark Design

San Francisco CA

Nancy Hunt Weber

Everett Graphics

Heritage Kitchen Specialty Foods

Hornall Anderson Design Works, Inc.
Seattle WA
Jack Anderson
Julie Tanagi-Lock
Julia LaPine
Starbucks Coffee Co.

William Plewes Design Inc.
Toronto Canada
Chris Plewes
Gary White
Campbell's Soup Co.

Hermsen Design Associates, Inc.
Dallas TX
Jack Hermsen
James Sheehan
Quaker Oats Co.

Roy Ritola, Inc
San Francisco CA
Roy Ritola
Jeffrey Kaplan
Aztek, Inc.

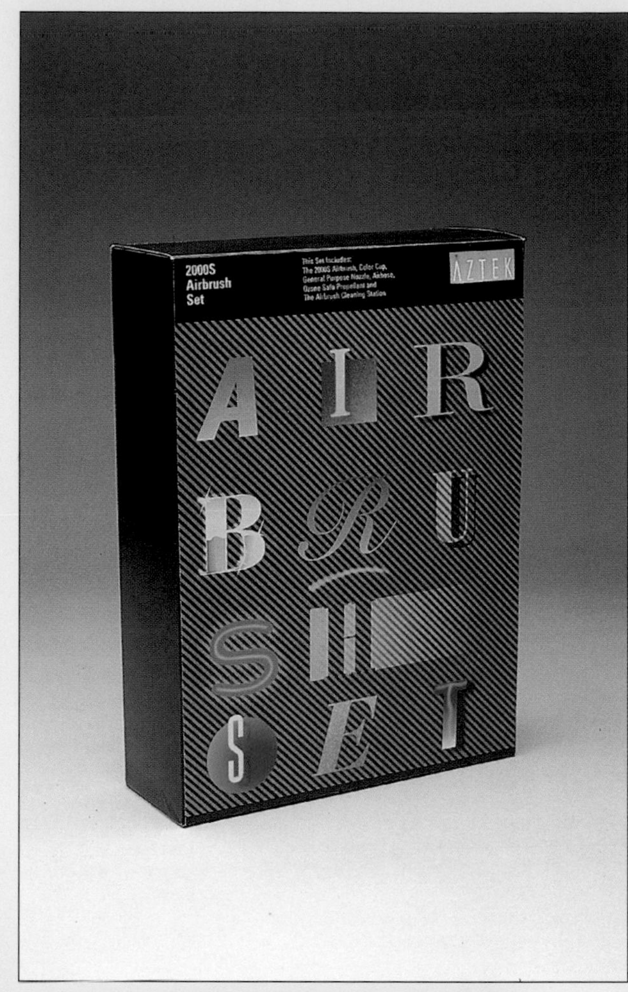

Mike Quon Design Office
New York NY
Mike Quon
Jack Anesh
Pressman Toys

B.E.P. Design Group
Brussels Belgium
Brigitte Evrard
Carole Purnelle
Laiteries Saint Hubert-France

PROFILE DESIGN
San Francisco CA
KENICHI NISHIWAKI
BRIAN JACOBSON
ANTHONY LUK
JEANNE NAMKUNG
Sanyo Department Store

Desgrippes Cato

Gobe Group/Paris

Paris France

Joel Desgrippes

Sophie Farhi

Beiersdorf Nivea

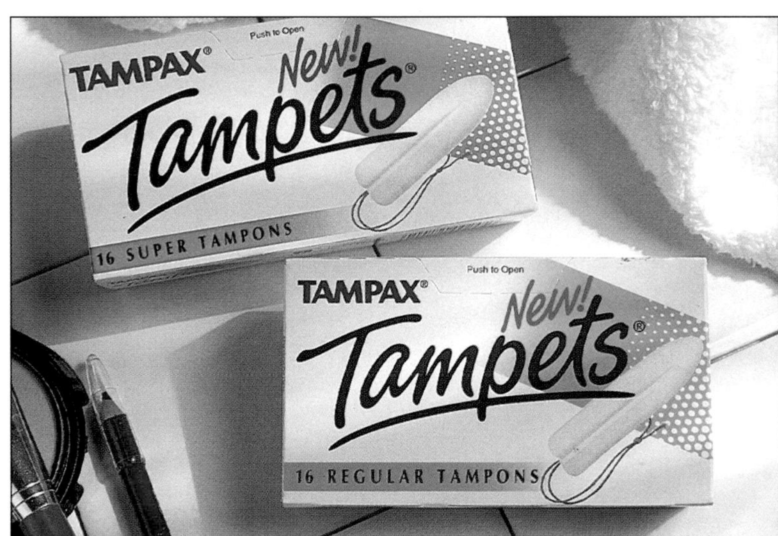

Coleman, LiPuma, Segal & Morrill, Inc.

New York, NY

Owen W.Coleman

Stephen H. Merry

Anthony Parisi

Linda Miller

Tambrands, Ltd. UK

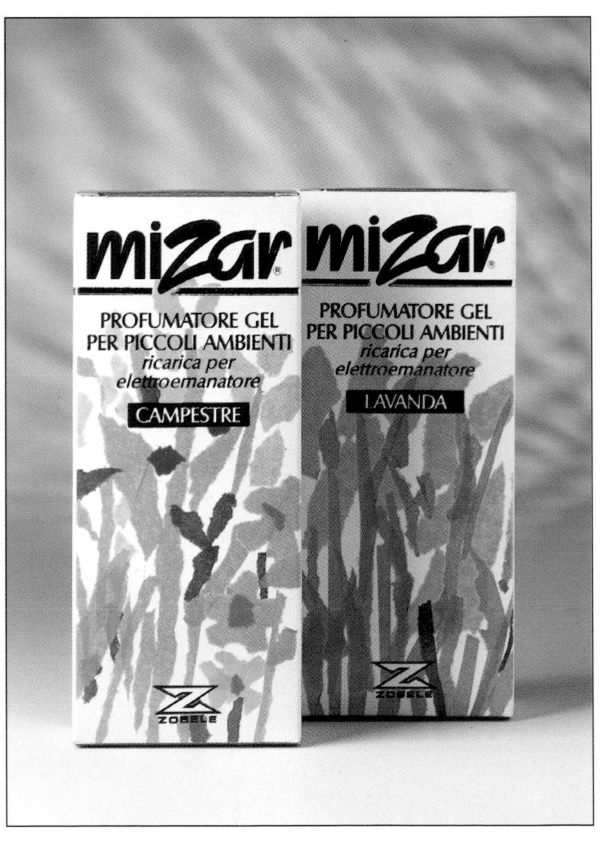

Design Group Italia

Milano Italy

Julien Beaheghel

Knut Hartmann

John Parsons

Zobele

Desgrippes Cato

Gobe Group/Paris

Paris France

Joel Descgrippes

Sophie Farhi

Sylvie Verdier

Lalique

Colonna Farrell: Design Associates

St. Helena CA

Ralph Colonna

Tony Auston

Tom Decker

Canadian Arctic Beverage Corp.

Moonink Communications
Chicago IL
Thomas Jones
John Downs
Bruce D. Beck
Rust-Oleum Corp.

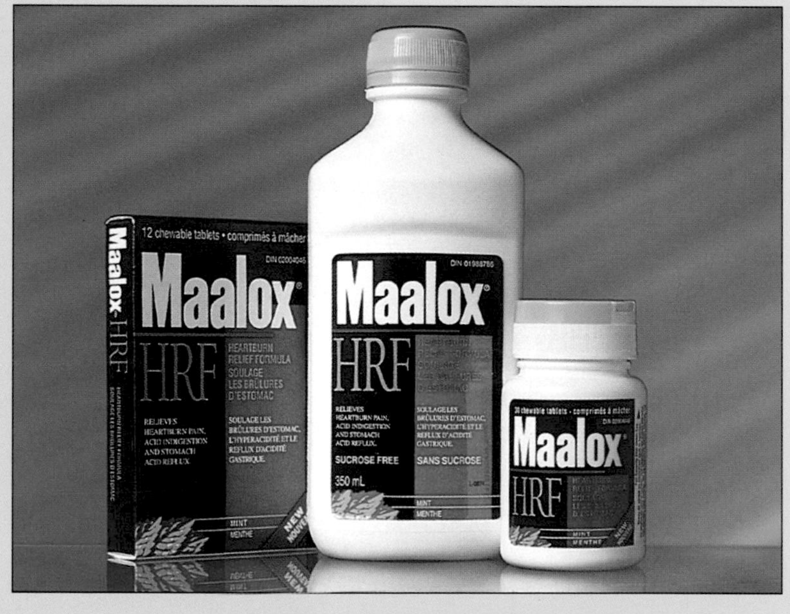

William Plewes Design Inc.
Toronto Canada
Chris Plewes
Rhone-Phoulenc Rorer

Graphique Design
Middlesex UK
Ray Armes
Barry Seal
Lever Brothers

Package Design of America
Bridgeport CT
Alan Anderson
Gary Holda
Ed Harrington
Bausch & Lomb

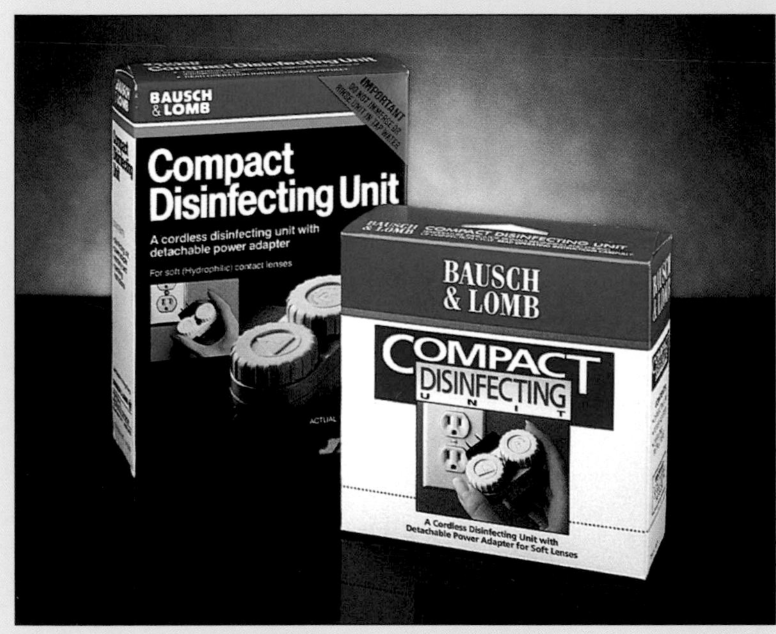

Forward Design, Inc.
Rochester NY
W. James Forward
Daphene Poulin
Karen Kall
Bausch & Lomb

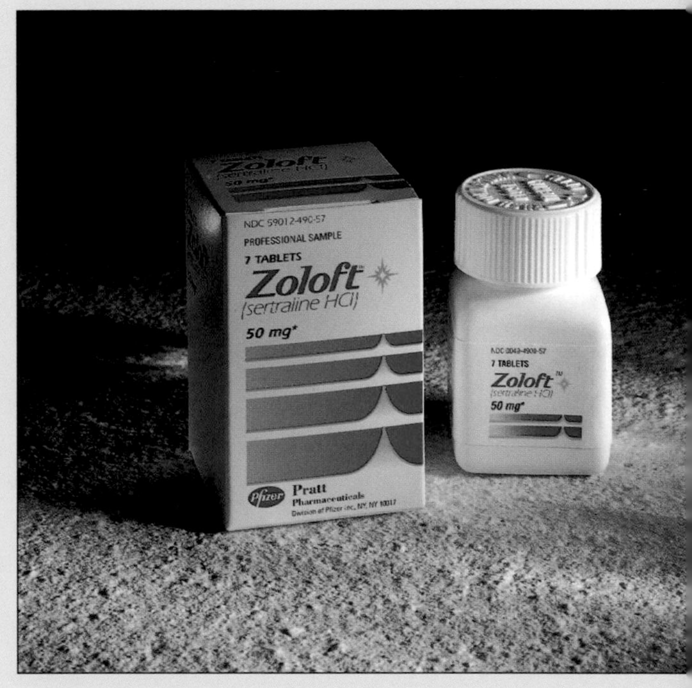

The Benchmark Group
Westport CT
Janice Jaworski
Vincent Massotta
Pfizer (Zoloft)

Wallner Harbauer Bruce

Chicago IL

Dwight O. Nelson Jr.

Joel Freeze

Charles Gillis

Schwinn Bicycles

The Van Noy Group

Torrance CA

Jim Van Noy

Dave Sapp

Joe Huizar

Kwikset Corporation

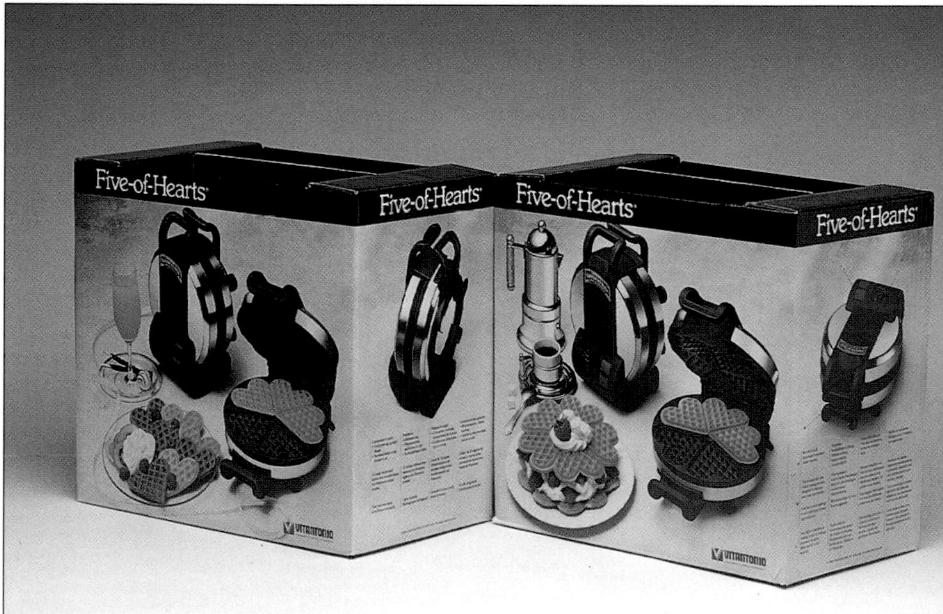

Karen Skunta & Company

Cleveland OH

Karen A. Skunta

Beth Segal

Elayne Lakrtiz

Vitantonio Mfg. Co.

Goldsmith Yamasaki Specht Inc.
Chicago IL
Claude Cummings
ITT Aimco

Deskey Associates
New York NY
Chris Merwin
Scott Yaw
Ellen Foreman
Black & Decker/DeWalt

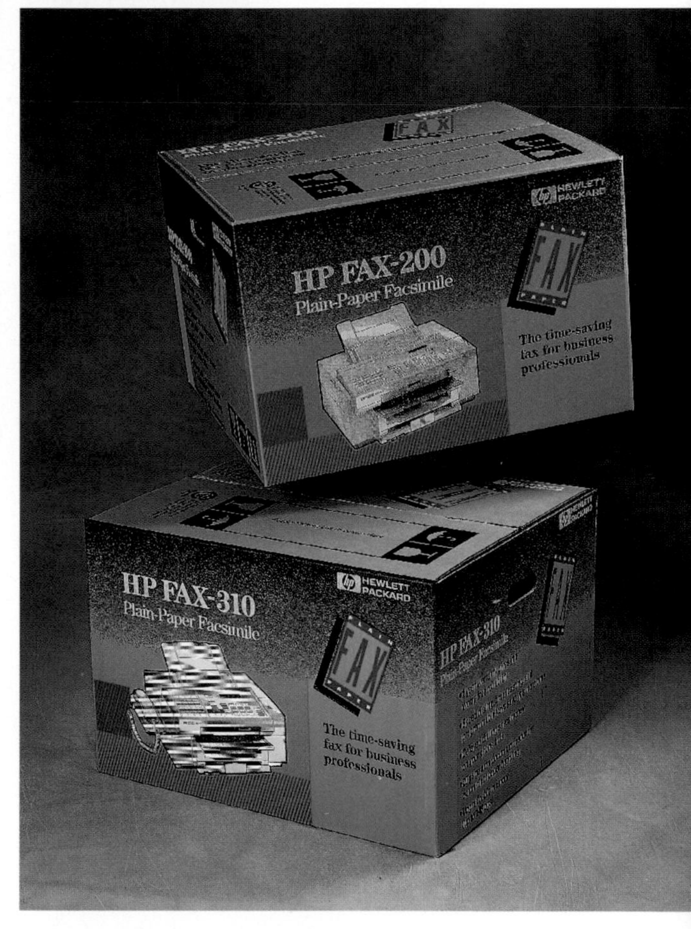

Hewlett-Packard Company
San Diego CA
Alan Klay
San Diego Division

Nabisco Foods, Inc.
Parsippany NJ
Sterling Design
Elizabeth Morales
George Hoffman
Nabisco Foods, Inc.

Tim Girvin Design, Inc.
Seattle WA
Tom James/Nabisco
Tim Girvin
Karl Leuthold
Bevin Perrine
Nabisco Foods Grp.

Design Forum
Dayton OH
Bruce Dybvad
Carolyn Zudell
Diane Acker
Del Taco

Nabisco Foods, Inc.
Parsippany NJ
Wallace Church
Associates
Lesley Tucker
Angela Valenti
George Hoffmann
Nabisco Foods, Inc.

Yao Design International, Inc.
Tokyo Japan
Yoshiaki Koma
Takeo Yao
Koji Hasegawa
Ezaki Glico Co., Ltd.

The Thomas Pigeon Design Group
Toronto Canada
The Thomas Pigeon
Design Group
Kraft General Foods

Supon Design Group, Inc.
Washington DC
Supon Phornirunlit
Richard Lee Heffner
Madison Square Press
Iconopolis

Lipson-Alport-Glass & Associates
Northbrook IL
Sam J. Ciulla
Ann Werner
Gary D. Majus
International Games, Inc.

Libby Perszyk Kathman
Cincinnati OH
Braddly J. Bush
Tina Fritsche
Totes

HENRIK OLSEN
Pasadena CA
HENRICK OLSEN
Second Hand Athlete

WALLACE CHURCH ASSOCIATES, INC.
New York NY
STANLEY CHURCH
JOE CUTICONE
Shaw Nautical

Nick Lane Design
Sausalito CA
Nick Lane
Anne Chambers
Brian Leatart
Olde Thompson

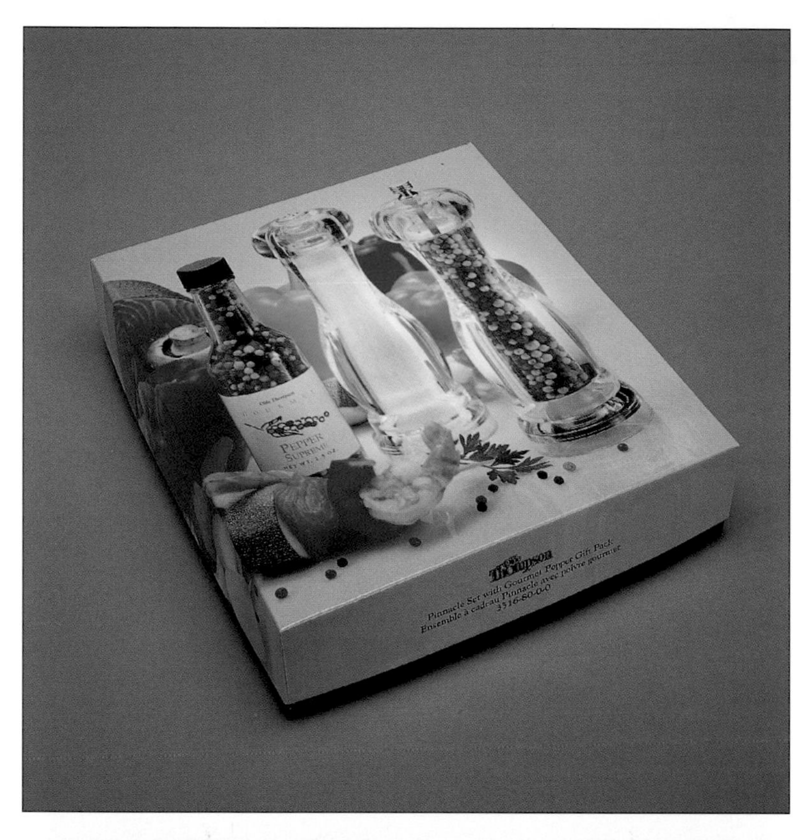

Fatta Design Group, Inc.
Greenwich CT
Pattie Gabberty
Joe De Jesus
Fatta Design Group
Avis Rent A Car

Yonetsu Design House
Tokyo Japan
Hisakichi Yonetsu
Bankaku Sohonpo Co., Inc.

Gerstman + Meyers Inc.
New York NY
Richard Gerstman
Rafael Feliciano
G+M Staff
American Standard Inc.

Georges Gotlib, Inc,
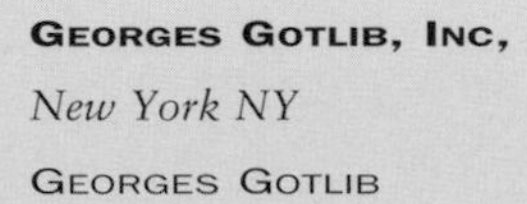
New York NY
Georges Gotlib
Julie Aronowitz
Michele Ranieri
Danskin, Inc.

Kollberg/Johnson Associates
New York NY
Penny Johnson
Gary Kollberg
Arthur Wang
Corning Vitro

Design North, Inc.
Racine WI
MArk Topczewski
Jim Wend
BWD Automotive Corporation

Harrisberger Creative
Virginia Beach VA
Lynn Harrisberger
Claire Lingenfelser
Barker, Campbell & Farley
Harmony Products Canada, Inc.

Murrie, Leinhart, Rysner & Associates
Chicago IL
Kate McSherry
Kaytee

The BoardRoom Design Group
Cleveland OH
Kim Zarney
Bob Wood
Allan Snider
KenVet Veterinary Nutritional Products

ID exposure, Inc.
Wichita KS
Andrew J. Covault
Nicole Dermikaelian
Cindy Wells
Scott Luty
The Coleman Company, Inc./International Division

B.E.P. DESIGN GROUP

Brussels Belgium

GEORGE ROSENFELD

BRIGITTE EVRARD

CAROLE PURNELLE

Karlsberg Brauerei

Design Partners
Racine WI
Hollie Powell
Jim Jedlicka
Wendy Patty
David
Willardson/Bright
& Associates
Miller Brewing Co.

Suntory Limited
Osaka Japan
Hiroshi Yokoo
Katsuyuki Fujimoto
Tomosada Ukegawa, Studio Wild
Suntory Limited

D'Addario Design Associates, Inc.
New York NY
Adam J. D'Addario
D.G. Yuengling & Son, Inc.

TAB Graphics
Denver CO
TAB Design Staff
Packaging Corporation of America
Aspen Pet Products

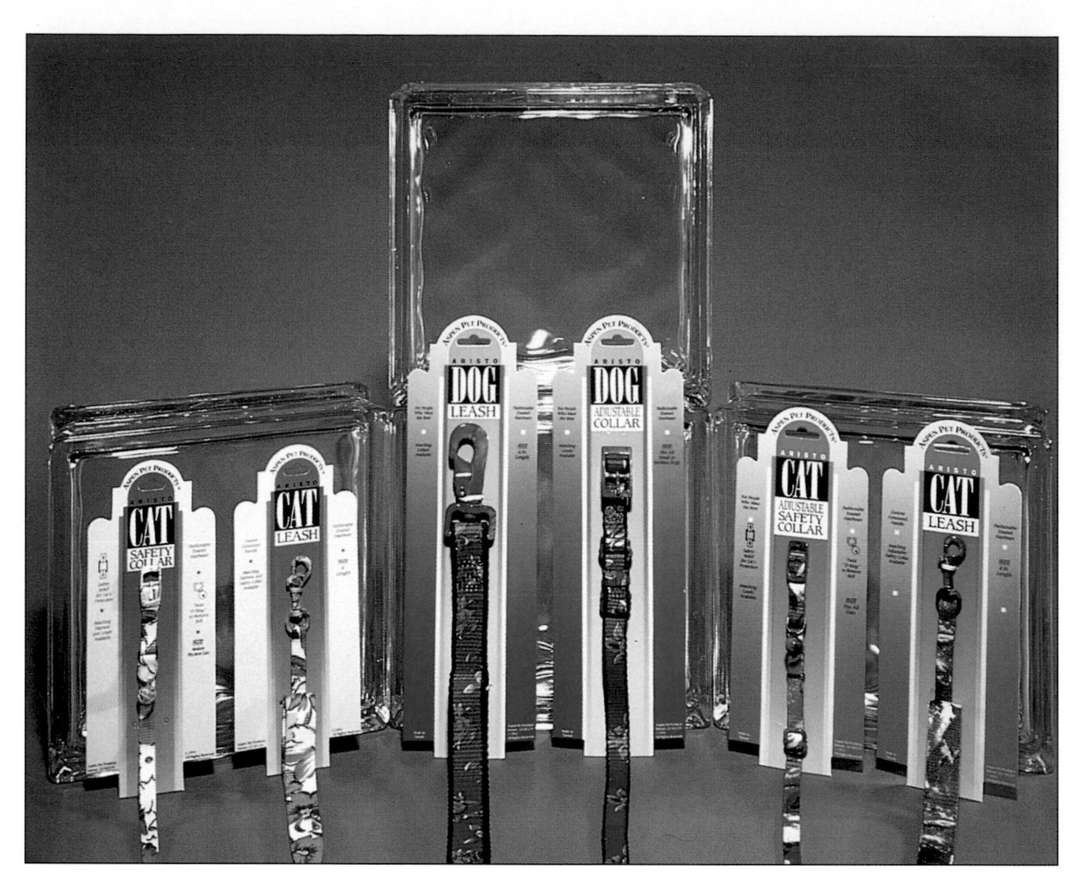

Sterling Design
NewYork NY
Heather Armstrong
Mike Bainbridge
The House of Seagram

Revlon, Inc.
New York NY
William Lunderman
Bonnie Moore
Corwin
David Jones
Quik Takes

Suntory Limited

Osaka Japan

Toshihiko Daimon

Yoshio Kato

Yukiko Shibato

Tomosada Ukegawa, Studio Wild

Suntory Limited

Source/Inc.

Chicago IL

Hedy Lim Wong

Deborah McKibben

McCain Citrus Inc.

Peterson & Blyth Associates
New York NY
Alex Pennington
The American Tobacco Company

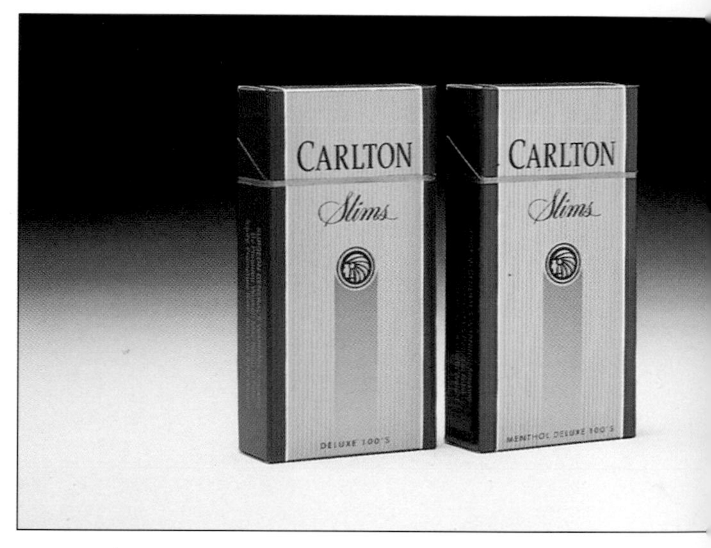

Jager DiPaola Kemp Design
Burlington VT
Michael Jager
David Covell
Adam Levite
XXX Snowboard Wax

Mittleman/Robinson Inc.
New York NY
Fred Mittleman
Carrie Friedman
Canandaigua Wine Company

Klaus Wuttke & Partners Limited
London UK
Klaus Wuttke
Liz Gutteridge
Cussons (UK) Ltd.

Gerstman & Meyers Inc.
New York NY
Michael Lucas
Lisa Keyko
Gloria Ruenitz
Carter Wallace International

Werbin Associates Inc.
Mamaroneck NY
Xavier De Eizaguirre
Baroness Philippine De Rothschild
Baron Philippe De Rothschild S.A.

Curtis Design
San Francisco CA
Justine Carroll
David Curtis
Rick Jansen
Granny Goose Foods

Hartmann & Mehler Designers GmbH
Frankfurt Germany
Dunja Nonnenmacher
Walter Pepperle
Graham Boyes
Hans-Jurgen Mages
Donna Grant
Asbach GmbH & Co.

Barrett Design Inc.
Cambridge MA
Karen Dandy
Anne Callahan
Mathsoft Inc.

Britton Design
Sonoma CA
Patti Britton
Sam Sebastiani
Jesse Kimin
Viansa Winery

Miller Sutherland
London UK
Kathy Miller
David Stewart
Waitrose

Mark Oliver, Inc.
Santa Barbara CA
Mark Oliver
Susan Edelman
Jay Boettner
Adrienne's Gourmet Food

Hillis Mackey & Co., Inc.
Minneapolis MN
Tom Roper
Terry Mackey
Patrick Faricy
The Pillsbury Co.

Bronze

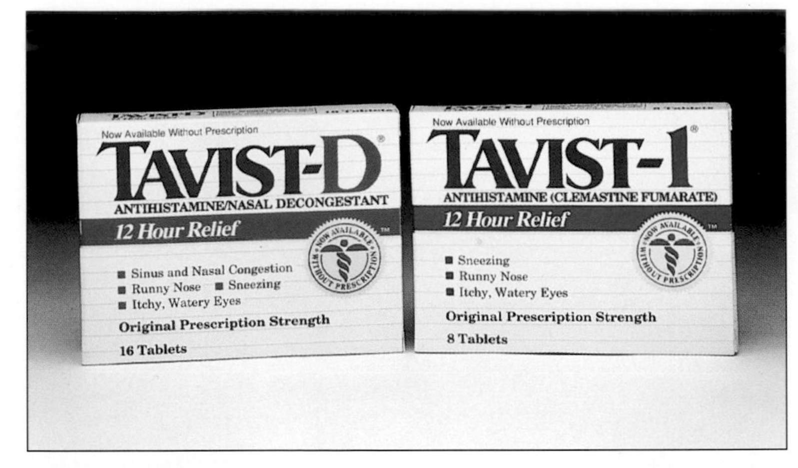

Peterson & Blyth Associates
New York NY
Janeen Violante
Ronald Peterson
Sandoz Pharmaceuticals Corp.

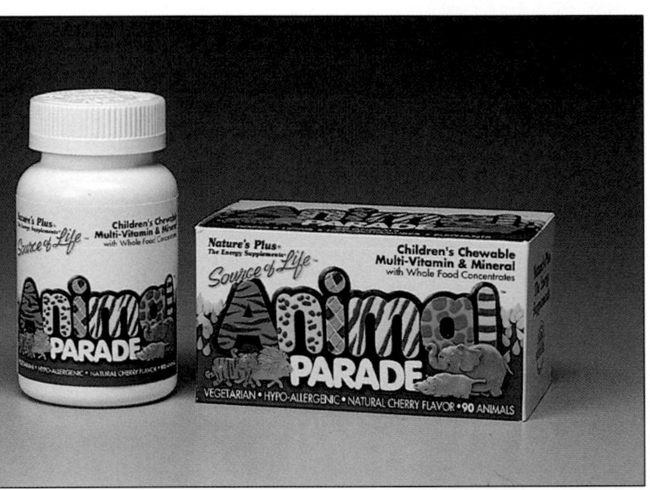

Apple DesignSource, Inc.
New York NY
Barry G. Seelig
Catherine Levine
Jacki Snider
Natural Organics, Inc.

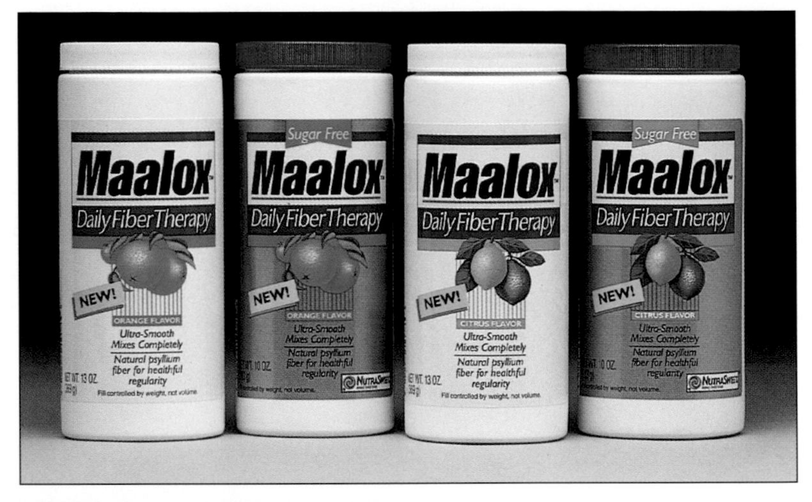

Deskey Associates
New York NY
Genie King
Scott Yaw
Dave Paragamian
Rhone-Poulenc Rorer

Kornick Lindsay
Chicago IL
Kornick Lindsay
Smith Kline Beecham

Gunn Associates

Boston MA

James M. Keeler

Sam Petrucci

Gillette

Baker Jazdzewski Ltd.

London UK

Andrzej Jazdzewski

Nikki Marechal

Point To Point

Jeyes Ltd.

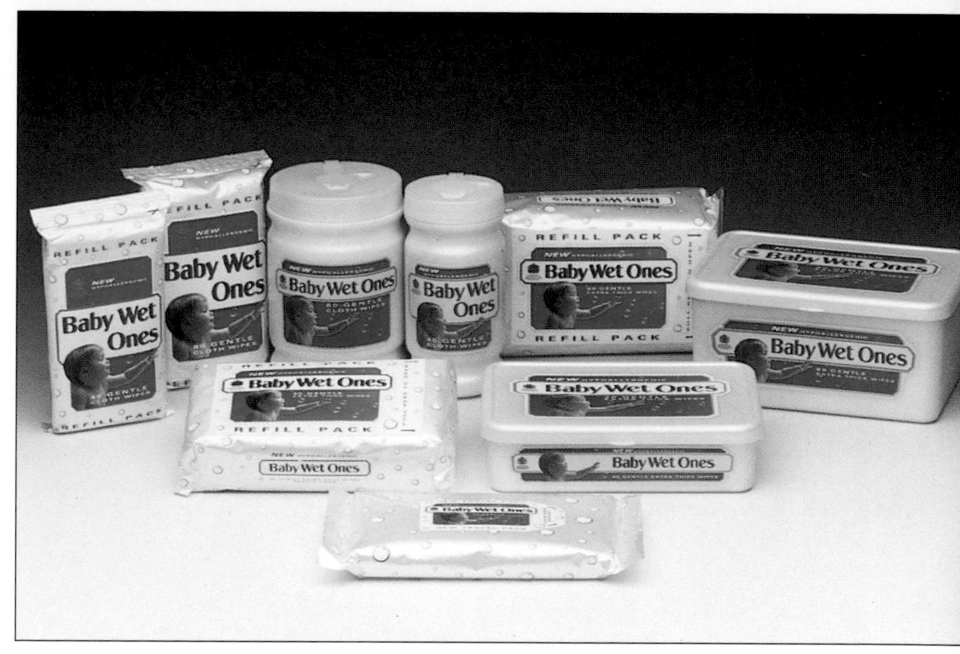

KETCHUM PROMOTIONS
New York NY
ARTHUR PALAZZO
JANEY BOYERONUS
SUSAN STULZ
MCI International, Inc.

SEAN MICHAEL EDWARDS DESIGN, INC.
New York NY
EDWARD O'HARA
LEAH HARTOG BRENMAN
Harvest Consulting, Inc.

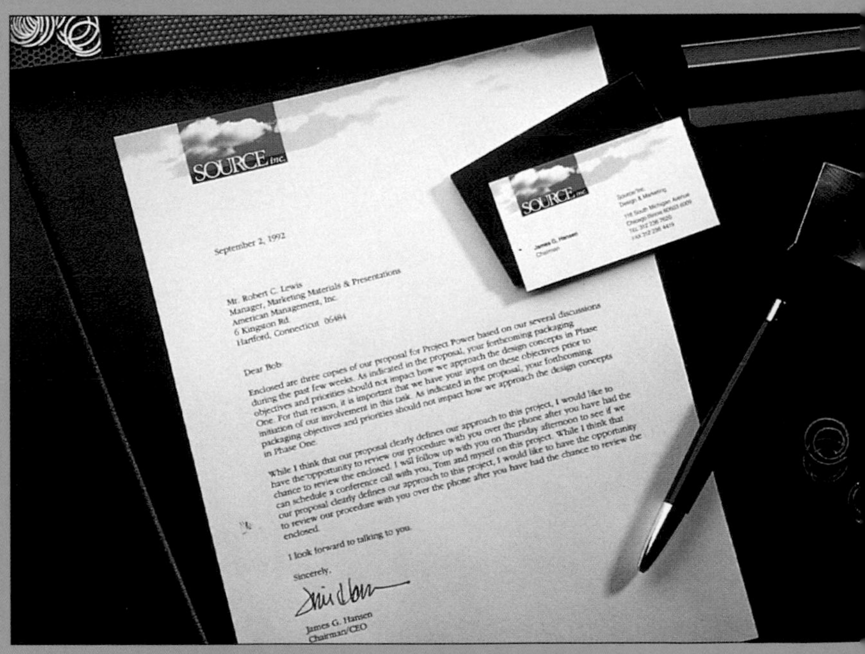

SOURCE/INC.
Chicago IL
MICHAEL LIVOLSI
BERNIE DOLPH
Source/Inc. Creative Staff.

MVP/Marketing, Visuals & Promotions, Inc.
Minneapolis MN
Greg Schultz
Brent Bentrott
Dick Weinrid
3M Consumer Stationery Division

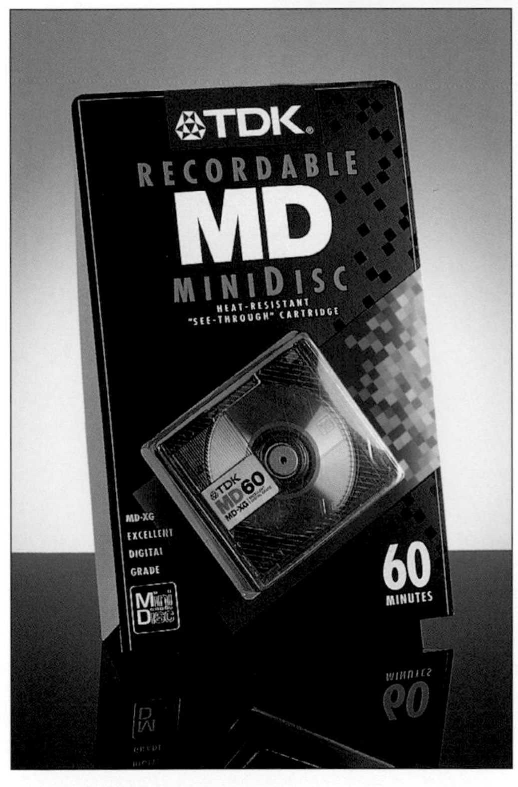

Handler Design Limited
White Plains NY
Bruce Handler
Rau Ringston III
TDK Electronics Corp.

Frankfurt Balkind Partners
New York NY
Aubrey Balkind
Kent Hunter
Andreas Combuchen
Adobe Systems Incorporated

DesignCenter of Cincinnati, Inc.
Cincinnati OH
Jim O'Brien
Julie Eubanks
Jim Makstaller
Kenner Products

Hasbro, Inc.
Pawtucket RI
Matt Lizak
Philip Arndt
David Desforges
Creative Services

DIL Consultants in Design
Sao Paulo Brazil
Batavo

Landor Associates
San Francisco CA
Quentin Murley
Jon Weden
The Coca-Cola Co.

Lindt & Sprungli (USA) Inc.
Stratham NH
Susan M. Wright
Joel Bronz
Lindt Chocolate

Coley Porter Bell
London UK
Stephen Franks
Angela Richards
Haulmark

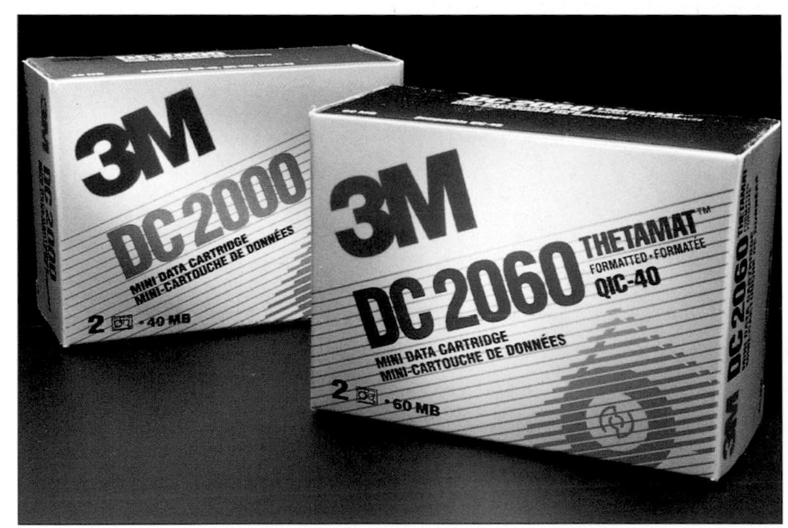

Frink Semmer and Associates, Inc.
Minneapolis MN
Sharon Sudman
Debbie Fiorella
Dan Deuel
3M Data Storage Markets Division

ABKCO Films
New York NY
ABKCO Films
Serino Coyne Inc.
Queens Design
Gimme Shelter/The Rolling Stones

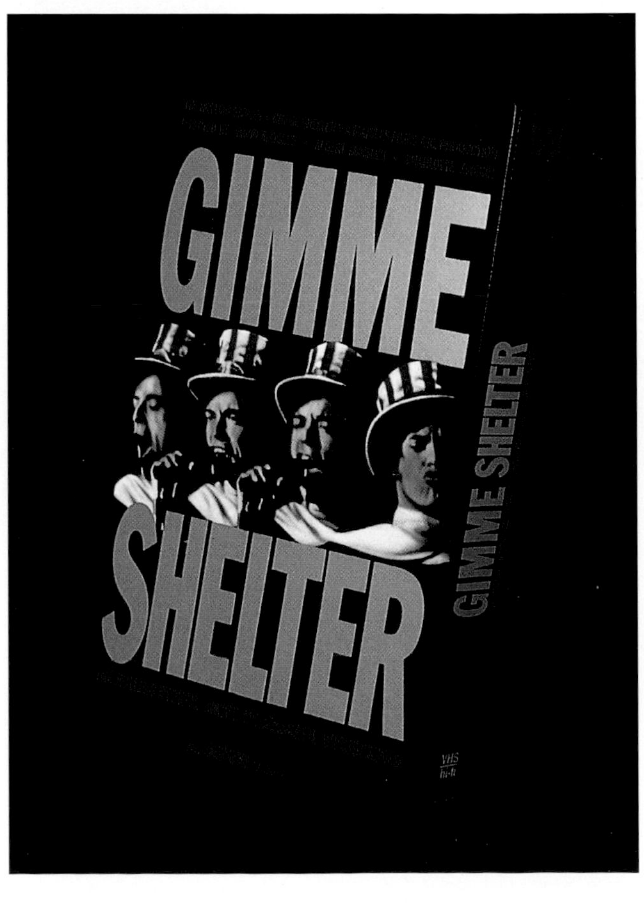

Primo Angeli Inc.
San Francisco CA
Primo Angeli
Rolando Rosler
Philippe Becker
Verifone Inc.

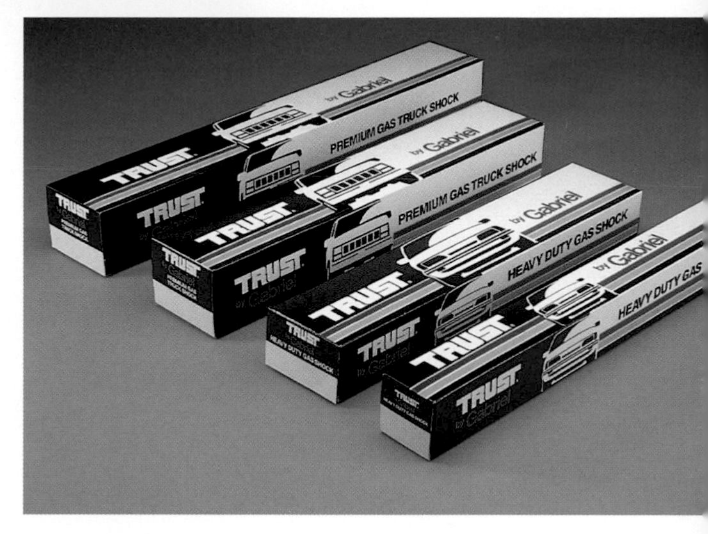

Biegler Design
Wheaton IL
Peter Biegler
Susan Gray
Gabriel Ride Control Products

Silver Burdett Ginn
Needham MA
Jodi Traub
Alice Hecht
Barbara Gazley/Hecht Design
Silver Burdett Ginn

The Great Atlantic & Pacific Tea Co., Inc.
Montvale NJ
Tony Lento
Karen Welman/Sterling Design
Michael Jackson/C.M. Jackson Co.
The Great Atlantic & Pacific Tea Co., Inc.

Murrie, Leinhart, Rysner & Associates
Chicago IL
Linda Voll
John Poll
Larry Bing
Linda Ware
Kraft General Foods

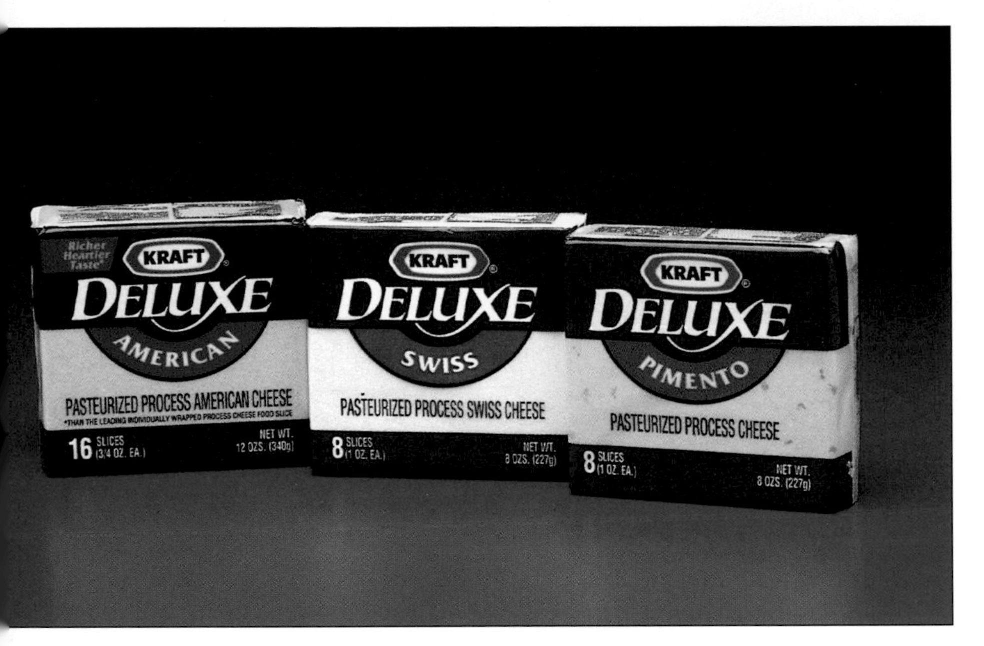

Wencel Hess Co.
Chicago IL
Jackie Lindert
Mike Wencel
Brian Banks
Jim Beam Brands Co.

Addison Design Consultants
San Francisco CA
Taco John's

Wallner Harbauer Bruce

Chicago IL

F. Jerry Harbauer

Steven Walker

Jay Highman

Worthington Foods, Inc.

Baker Jazdzewski Ltd.

London UK

Andrzej Jazdzewski

Gary Holt

Kim Lane

Point To Point

Courage Ltd.

Peter Seitz and Associates, Inc.
Golden Valley MN
Peter Seitz
Terri Hallman
Karn Knutson
Personnel Decisions, Inc.

Reed & Barton
Taunton MA
Ed Avadesian/Jewel Case
Reed & Barton

Tactix Communications & Design

Kitchener Canada

Synmark Marketing Strategies

Oakville, Canada

Tandet Nationalease Ltd.

Pethick and Money Limited

London UK

Julian Money

Kam Oswald

Alfred Dunhill Limited

MCI Design GmbH
Frankfurt Germany

Hughes Design, Inc.
Norwalk CT

DiDonato Associates

Chicago IL

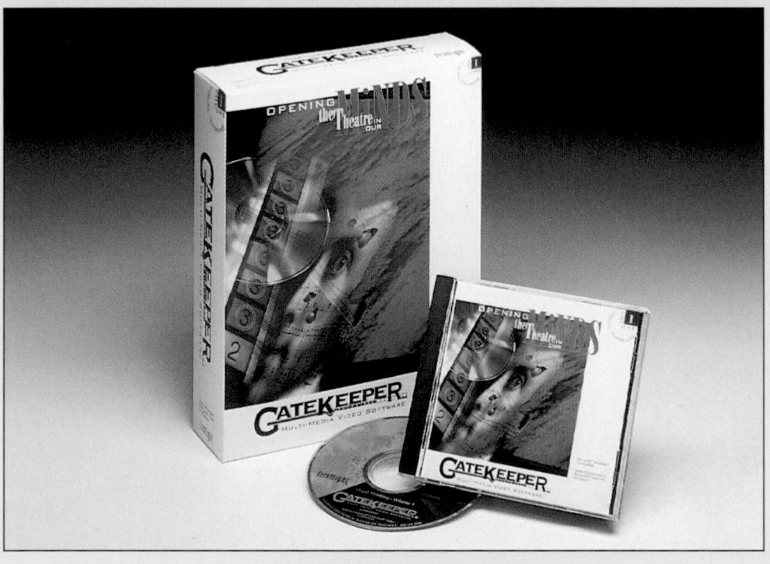

Ilium Associates, Inc.

Bellevue WA

Honorable

Clark/Linsky Design, Inc.

Charlestown MA

Marketing And

Culver City CA

Howard Blonder & Associates

Downey CA

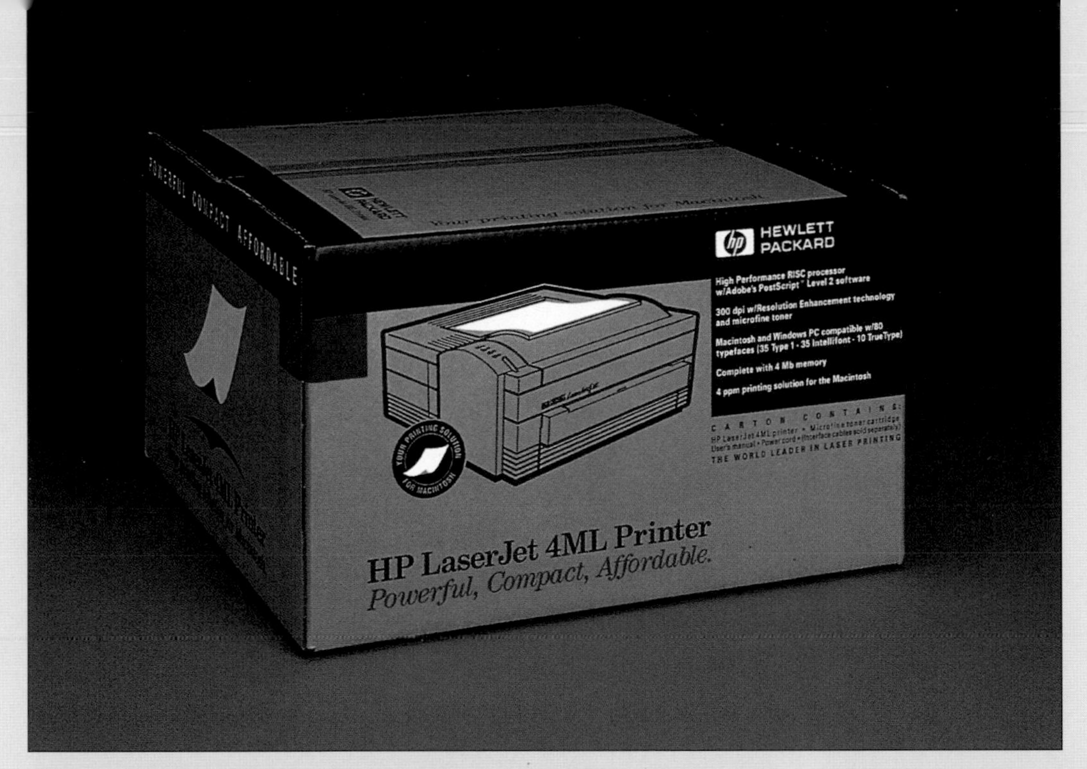

Hewlett-Packard Company
Boise ID

Besser Joseph Partners
Santa Monica CA

Mary Kay Cosmetics, Inc.
Dallas TX

The Weber Group Inc.
Racine WI

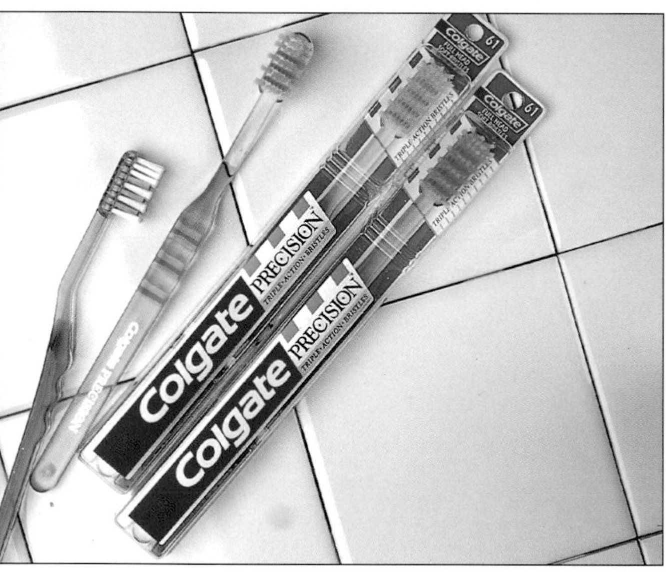

Colgate Palmolive Compnay
New York NY

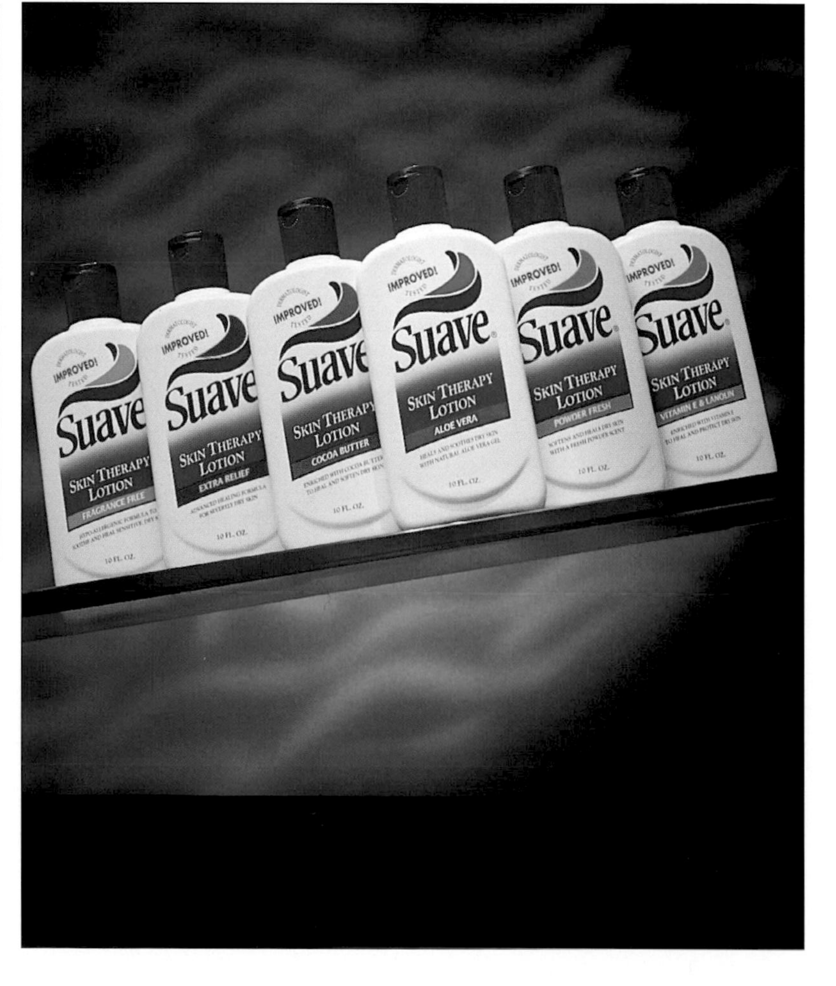

Helen Curtis, Inc.
Chicago IL

Hanson Associates, Inc.
Philadelphia PA

Lister Butler, Inc.
New York NY

Hood Design Group
Brookline MA

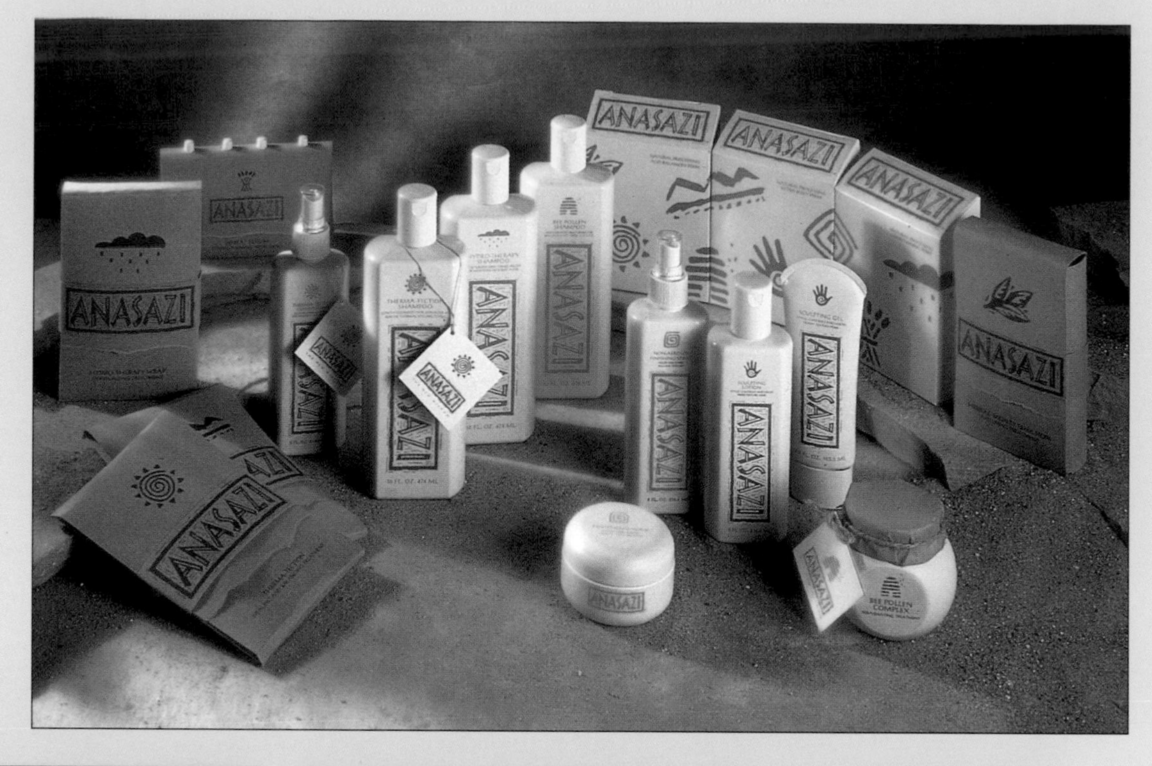

Michael Stanard, Inc.
Evanston IL
Anasazi

Hillis Mackey & Co., Inc.
Minneapolis MN
Target Douche

Fisher Ling & Bennion
Cheltenham UK
Calypso

MILLER-SUTHERLAND
London UK
Germaine De Cappucini

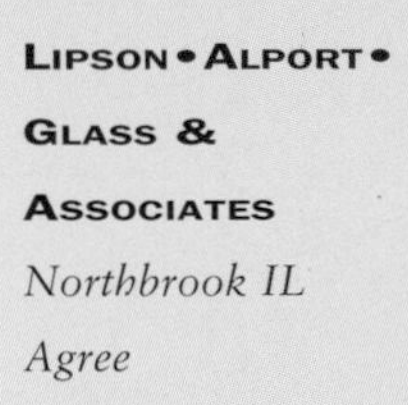

LIPSON • ALPORT • GLASS & ASSOCIATES
Northbrook IL
Agree

Verge LeBel Communications Inc.
Quebec Canada
Boulangerie

Tulocay Design Partnership
Napa CA
Trefethan Vineyards

The Benchmark Group
Westport CT
Classico

Source/Inc.
Chicago IL
Weber
Kanetetsu Delica Foods

Britton Design
Mill Valley CA
Riserva Anatra

Design Centre of Cincinnati
Cincinnati OH
Trauth

FISHER LING & BENNION
Cheltenham UK
Symonds Cider

THE LEONHARDT GROUP
Seattle WA
Elk Ridge

KOLLBERG/JOHNSON
New York NY
Christian Brothers

TDC/The Design Company
San Francisco CA
Corn Nuts

Robin Shepherd Studios
Jacksonville FL
Caribbean Condiment

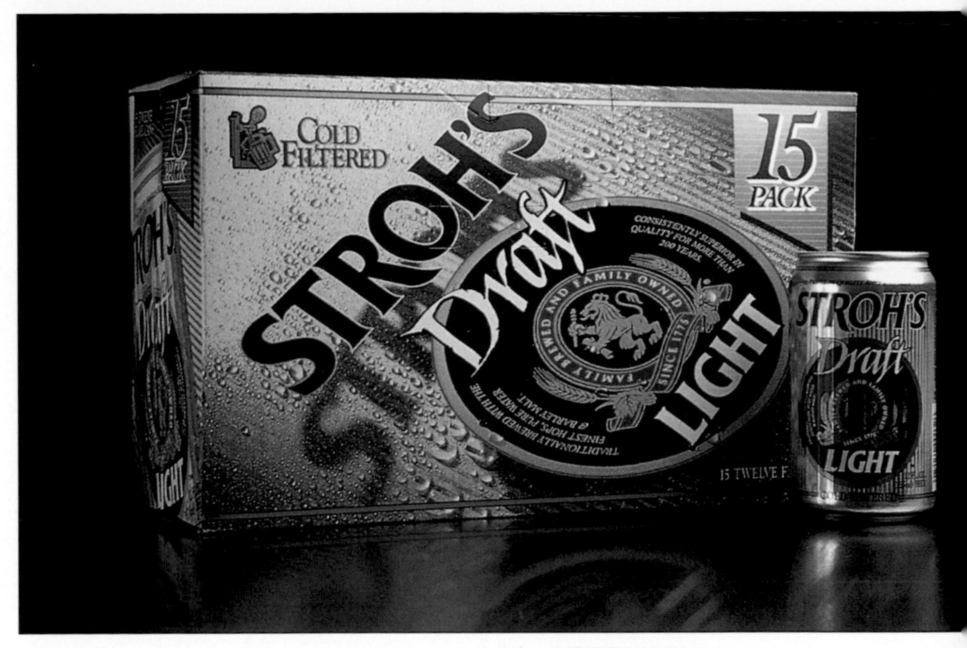

The Schechter Group
New York NY
Stroh's Draft

SUNTORY LIMITED

Osaka Japan

CHARON DYER DESIGN

Atlanta GA

Lullabelle's

PREPCO

Los Angeles CA

Toastettes

PROFILE DESIGN
San Francisco CA
Seadelica

SUNTORY LIMITED
Osaka Japan
Sparkling Wine Cocktail

THE THOMAS PIGEON DESIGN GROUP
Toronto, Ontario Canada
Dreamwhip

Libby Perszyk Kathman

Cincinnati OH

Totes Gloves

Biegler Design

Wheaton IL

Trust

Design Partners

Toronto, Ontario Canada

Bionaire

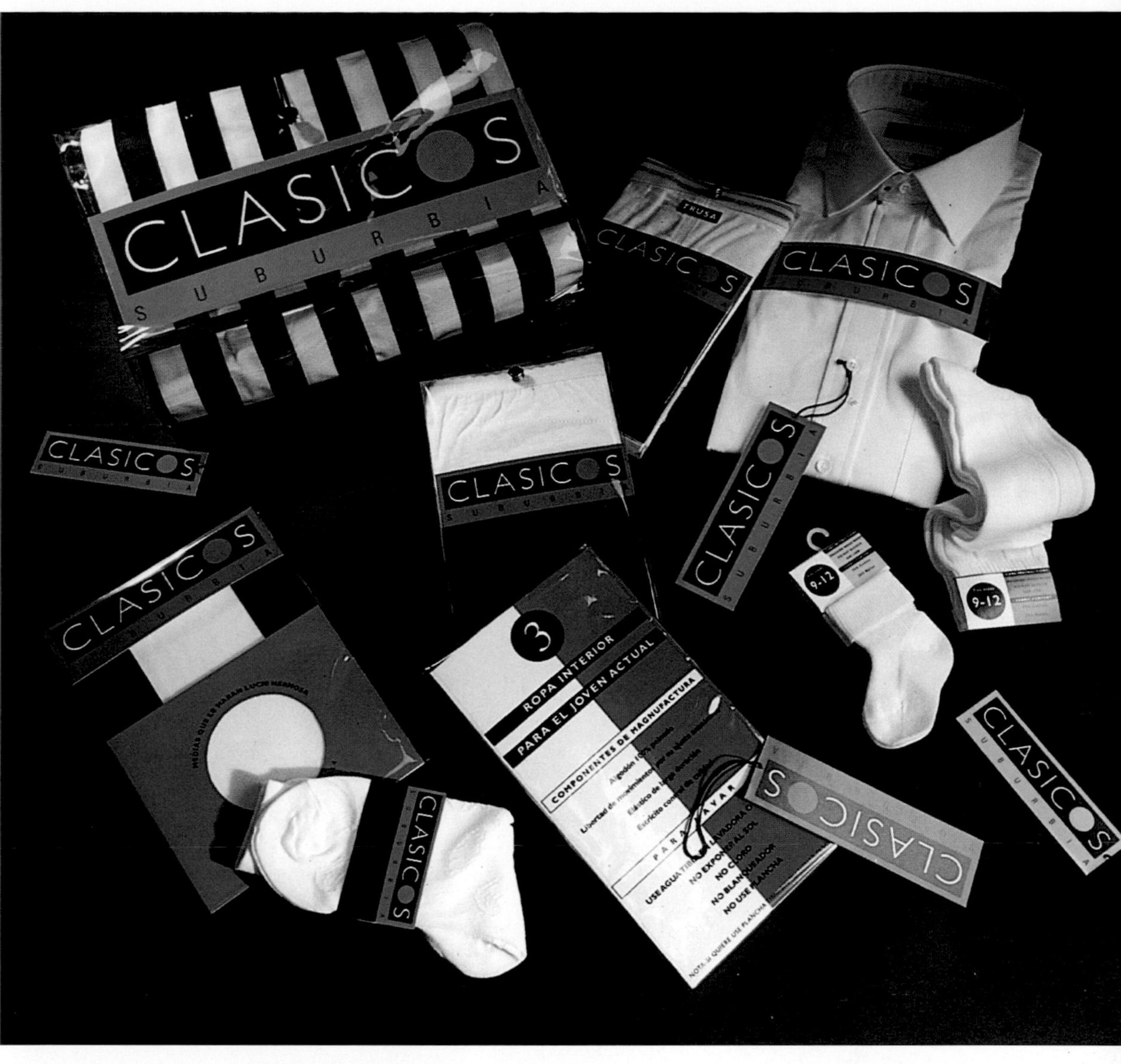

Schafer Associates, Inc.

Oak Brook Terrace IL

Classicos (Montage)

WBK Design

Cincinnati OH

Rockwell

DESIGN NORTH, INC.

Racine WI

Brite

DESIGN GROUP ITALIA

Milano Italy

Qubi

KORNICK LINDSAY

Chicago IL

Saran Wrap

Landor Associates
San Francisco CA
Montevina

Hunt Weber Clark Design
San Francisco CA
Effie Marie's

Design Bridge Ltd.
London UK
Oxo

Marketing, Visuals & Promotions
Minneapolis MN
Magic Crust

Curtis Design
San Francisco CA
Ausberger

Kornick Lindsay
Chicago IL
Sampoerna H & M Cigarettes

Besser Joseph Partners
Santa Monica CA
Roman Meal Lite

Design Partners
Toronto, Ontario
Canada
Heinz Batman

Elmwood
Guiseley, Leeds UK
Cabernet Merlot

Monnens-Addis Design
Emeryville CA
Dole

Gerstman & Meyers, Inc.
New York NY
Drixoral

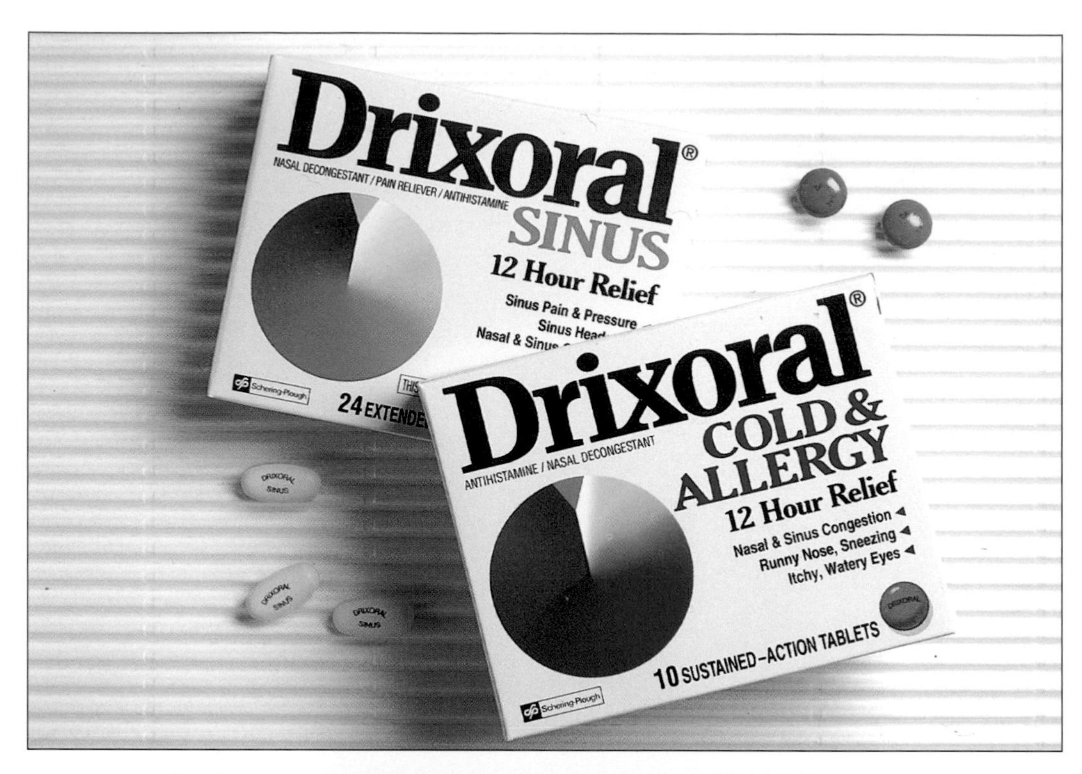

Coley Porter Bell
London UK
John Frieda

Hans Flink Design, Inc.
White Plains NY
NyQuil

MILLER-SUTHERLAND

London UK

Oils

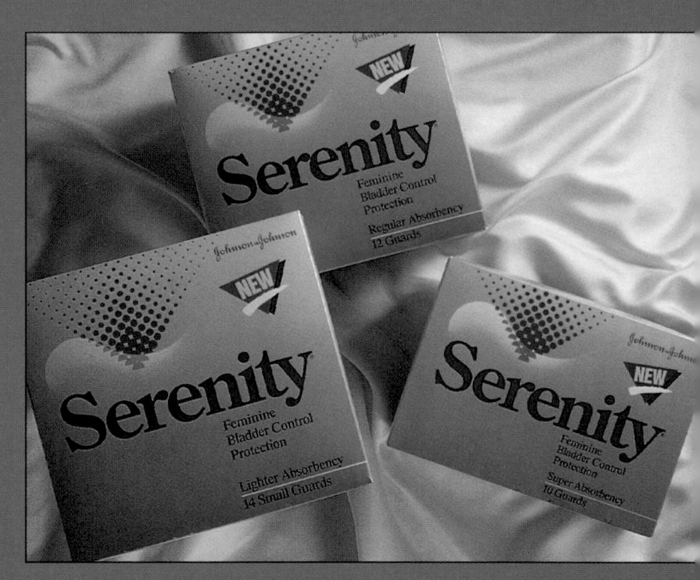

GERSTMAN & MEYERS, INC.

New York NY

Serenity

LIPSON • ALPORT • GLASS & ASSOCIATES

Northbrook IL

Bath & Body Works

The Benchmark Group

Westport CT

Bali

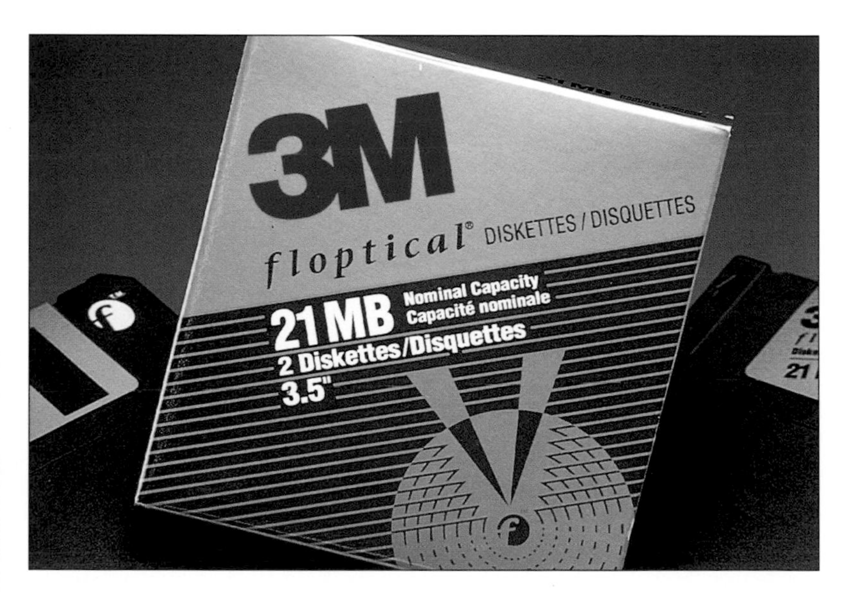

Frink Semmer & Associates Inc.

Minneapolis MN

3M Floptical

Tim Girvin Design, Inc.

Seattle WA

Math House

Design One
San Francisco CA
Seagate Flash Card

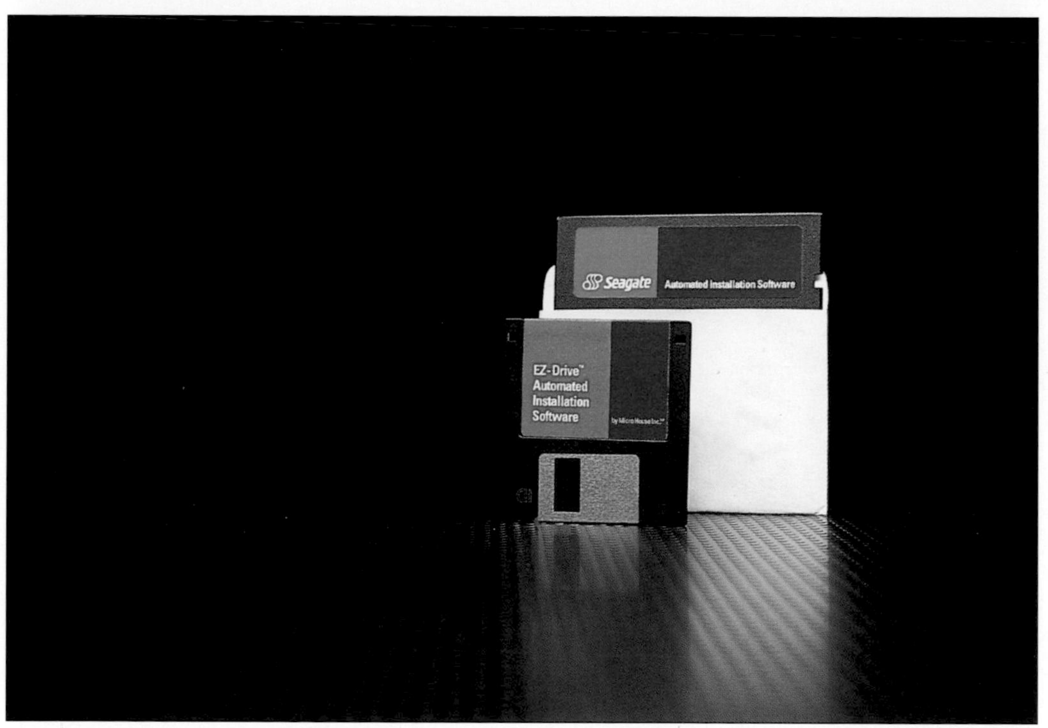

Addison Design
San Francisco CA
EZ-Drive

Primo Angeli, Inc.
San Francisco CA
Arch Rival

Special

Lewis Moberly
London UK
Tangs

B.E.P. Design Group
Brussels Belgium
Stabilac

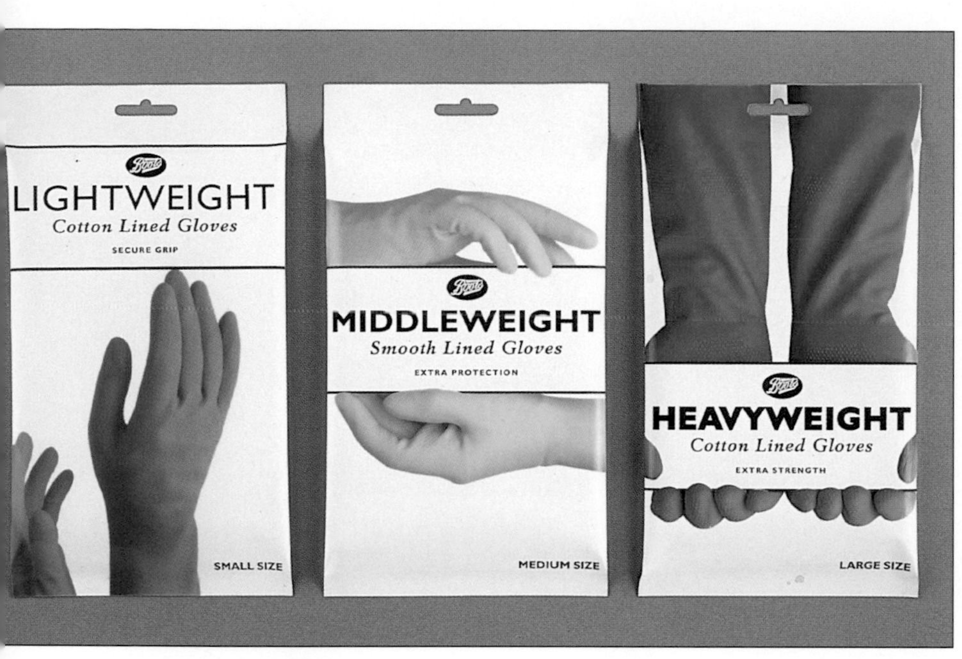

Lewis Moberly
London UK
Boots Household Gloves

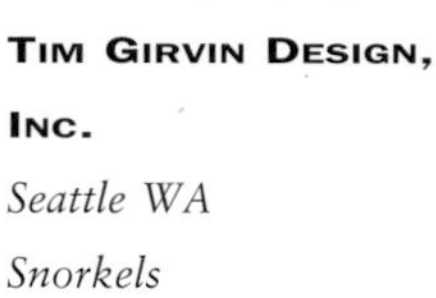

Tim Girvin Design, Inc.
Seattle WA
Snorkels

Source/Inc.
Chicago IL
Smuckers

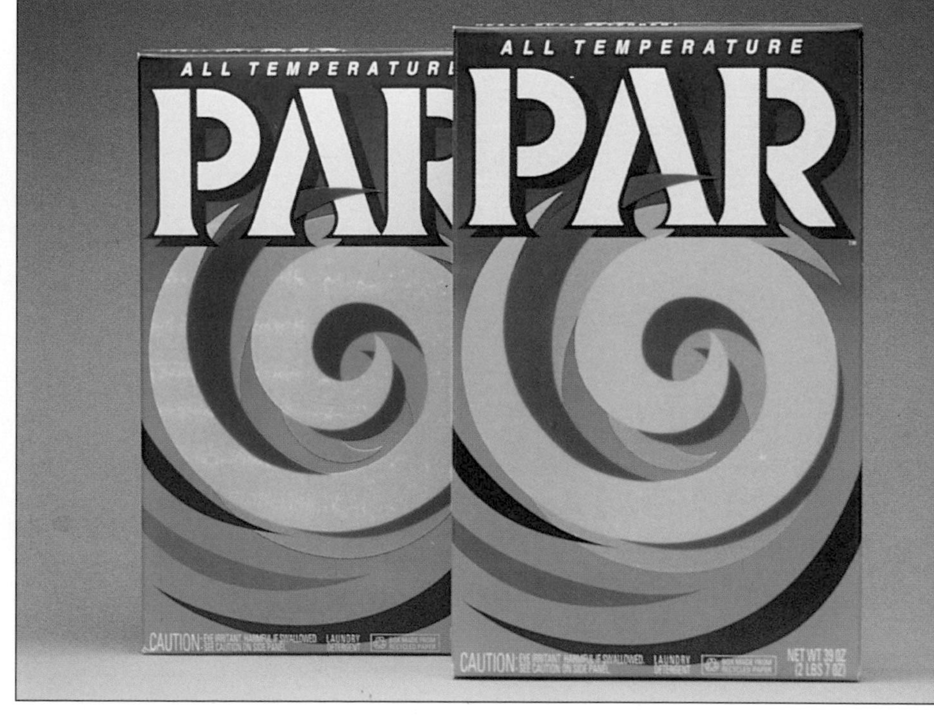

Primo Angeli, Inc.
San Francisco CA
PAR

D'Addario Design Associates, Inc.
New York NY
Andechs

TDC/The Design Company
San Francisco CA
DeKuyper

SUNTORY LIMITED

Osaka Japan

Suntory Pure Malt

WENCEL/HESS

Chicago IL

RonRico

Miller-Sutherland
London UK
Germaine De Cappucini

Lewis Moberly
London UK
Wizard

Colonna Farrell Design
St. Helena CA
VO5

LEWIS MOBERLY
London UK
Journey

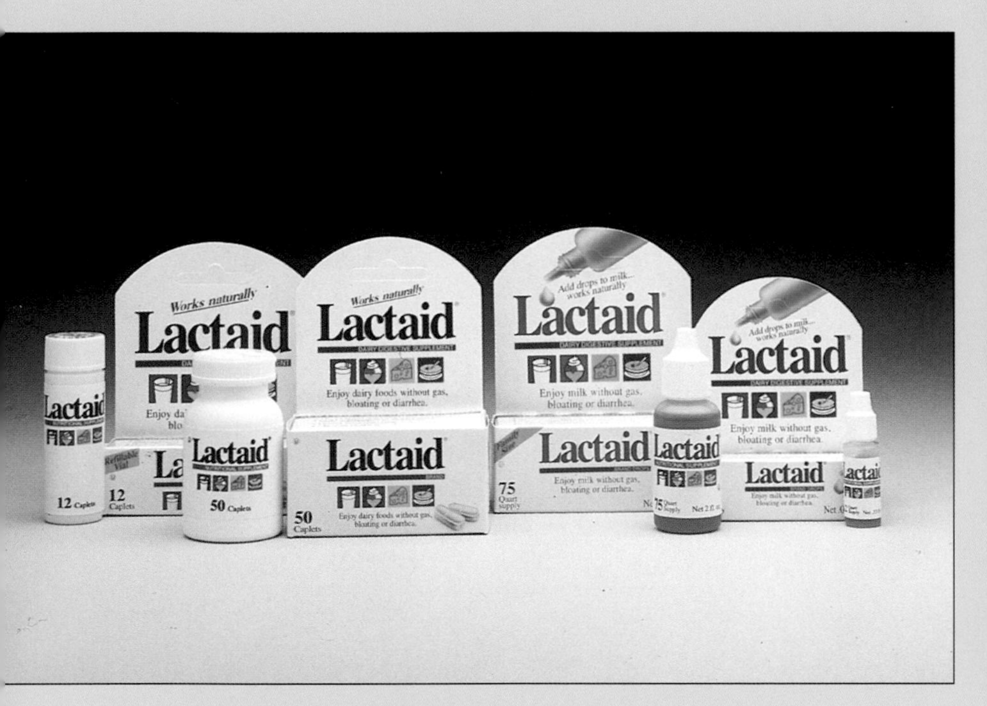

DAVID SCARLETT ASSOCIATES
New York NY
Lactaid

STERLING DESIGN
New York NY
Clarion

Coley Porter Bell

London UK

Brannigans

Kollberg/Johnson

New York NY

Planters Honey Roasted Peanuts

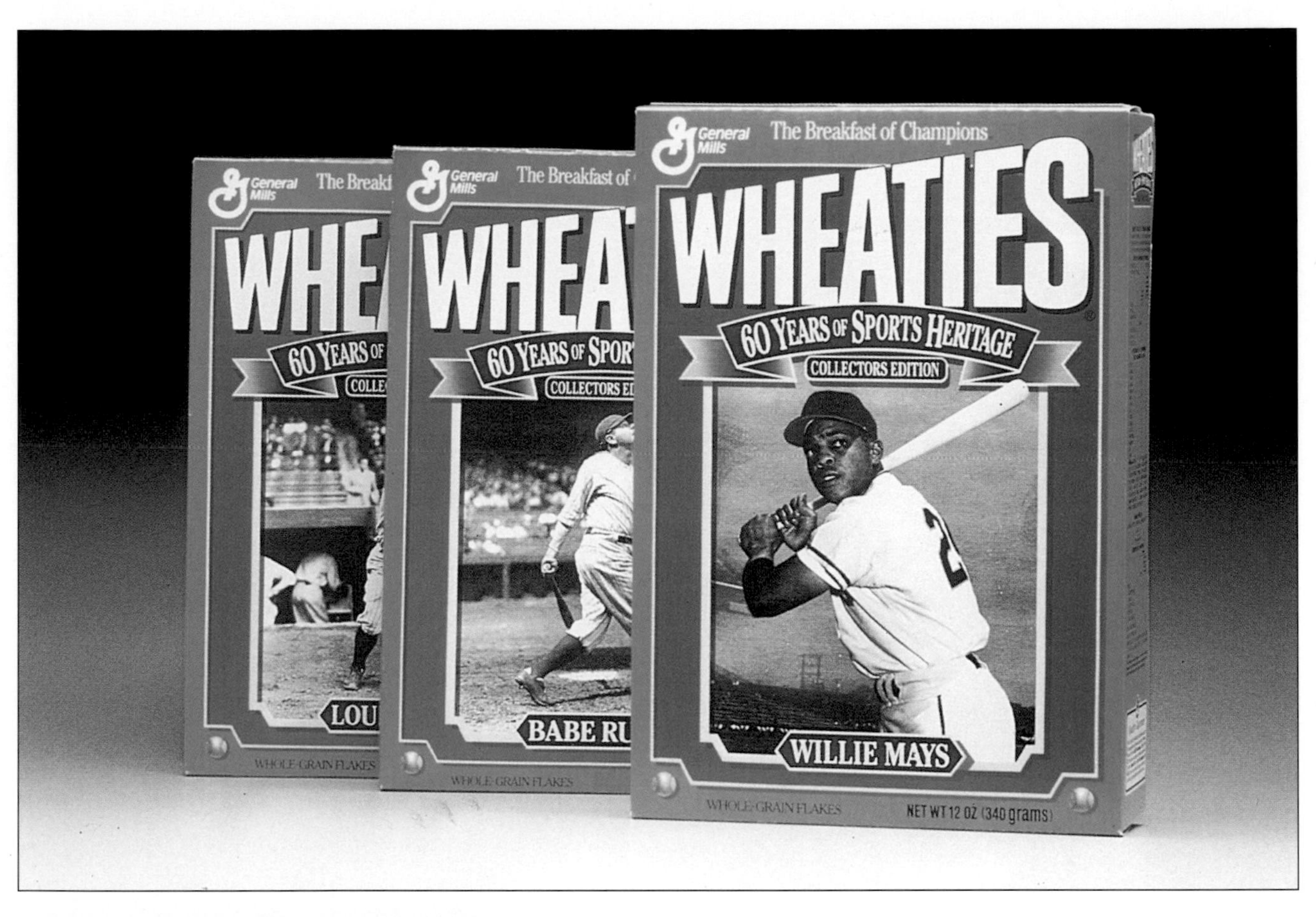

Hillis Mackey & Co., Inc.

Minneapolis MN

Wheaties

WBK Design

Cincinnati OH

Bob Evans

Besser Joseph Partners

Santa Monica CA

Roman Meal Original

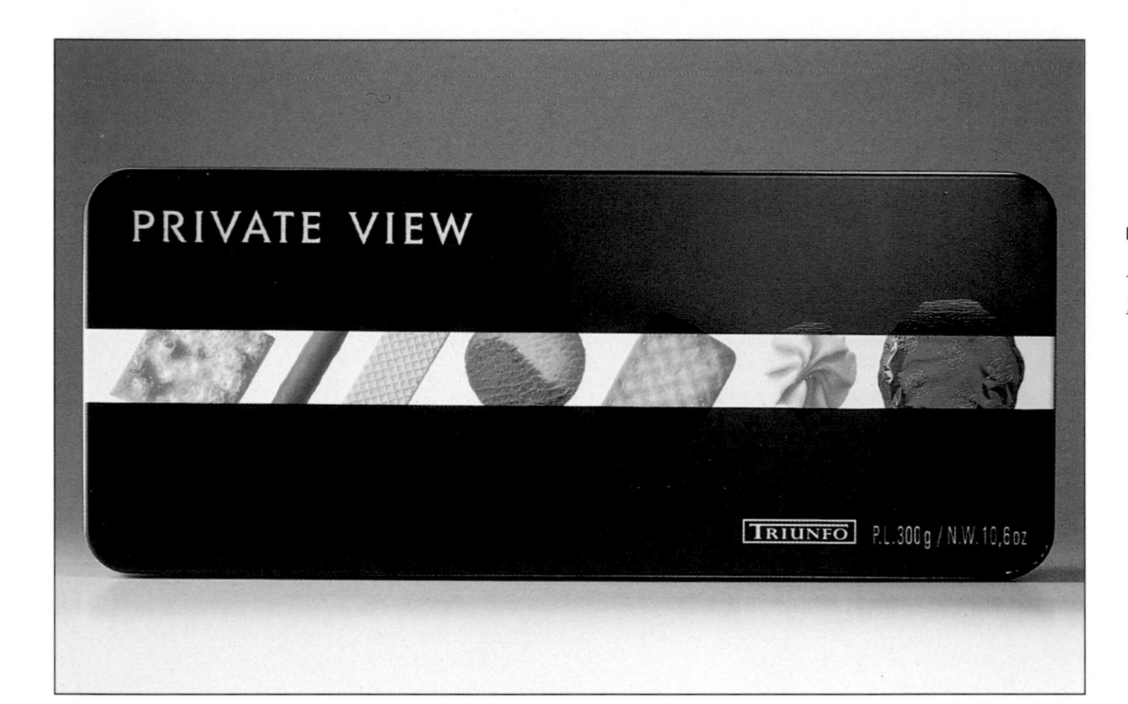

Lewis Moberly

London UK

Private View

Primo Angeli, Inc.

San Francisco CA

Mariani

Lewis Moberly

London UK

Duchy Originals

MONNENS-ADDIS DESIGN
Emeryville CA
Crust Cuisine

HARTE YAMASHITA & FOREST
Los Angeles CA
Knott's Light Preserves

THE GREEN HOUSE
London UK
Brown & Polson Chocolate Mousse

MILLEN & RANSON INC.
New York NY
Maple Valley

HARTE YAMASHITA & FOREST
Los Angeles CA
Knotts Extra Berries

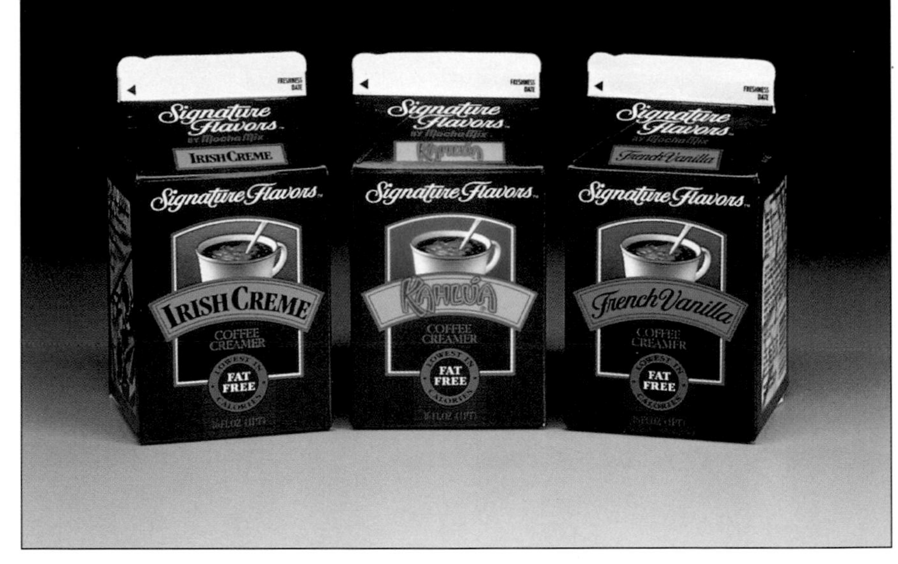

HARTE YAMASHITA & FOREST
Los Angeles CA
Signature Flavors

Lipson • Alport • Glass & Associates
Northbrook IL
Keebler Grahams

Hans Flink Design, Inc.
White Plains NY
Aero

Profile Design
San Francisco CA
Nature's Cupboard

PRIMO ANGELI, INC.
San Francisco CA
Italian Kitchen

SUPON DESIGN GROUP
Washington DC
Naturally

WILLIAM PLEWES DESIGN, INC.
Toronto, Ontario Canada
Le Menu

Hillis Mackey & Co., Inc.
Minneapolis MN
The Turkey Store

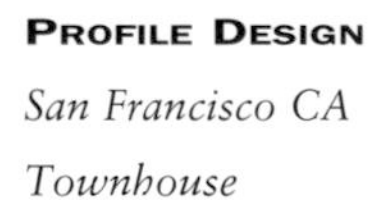

Profile Design
San Francisco CA
Townhouse

Gunn Associates

Boston MA

Right Guard

Special

Monnens-Addis Design

Emeryville CA

409

Charles Zunda Design Co., Inc.

Greenwich CT

Mr. Bubble

KOSE CORPORATION
Tokyo Japan
Cosmedecorte

KOLLBERG/JOHNSON
New York NY
Cutex

KOSE CORPORATION
Tokyo Japan
Intellige

Special

PRIMO ANGELI, INC.
San Francisco CA
Coors Cutter

COLONNA FARRELL DESIGN
St. Helena CA
Biale

COLEY PORTER BELL
London UK
Scottish Pride

Kornick Lindsay
Chicago IL
Coqui Malt Liquor

Sterling Design
New York NY
Miller Clear Beer

KRAFT GENERAL FOODS

Glenview IL

Velveeta

WALLNER HARBAUER BRUCE

Chicago IL

Ripple Crisp

WALLNER HARBAUER BRUCE

Chicago IL

Wheat Germ

WILLIAM PLEWES DESIGN, INC.
Toronto, Ontario
Canada
Fish Fries and Fillets

MOONINK COMMUNICATIONS
Chicago IL
Wonder Vienna

B.E.P. DESIGN GROUP
Brussels Belgium
Biologische Melk

DESIGN NORTH, INC.
Racine WI
Kix

LANDOR ASSOCIATES
San Francisco CA
Tropical

MORILLAS & ASOCIADOS
Barcelona Spain
La Lechera

KOLLBERG/JOHNSON
New York NY
Cocktails for Two

HILLIS MACKEY & CO., INC.
Minneapolis MN
Kemps Ice Cream

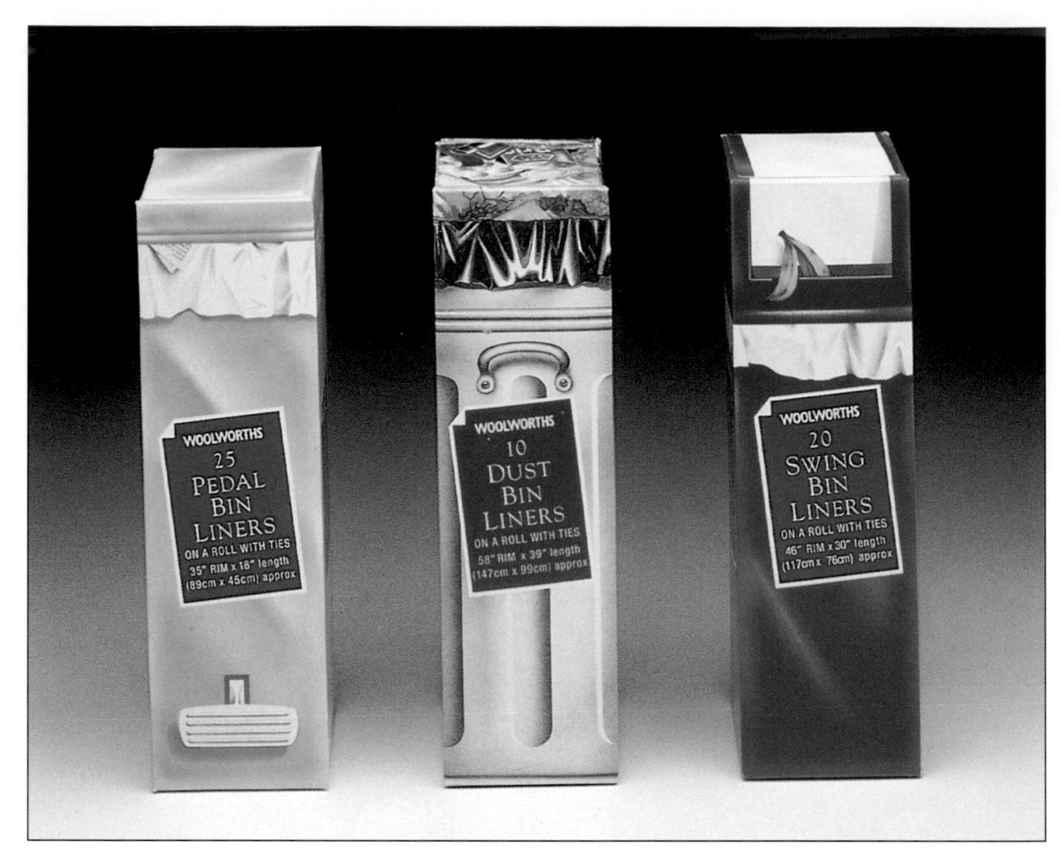

Coley Porter Bell
London UK
Woolworth's Bin Liners

Coleman, LiPuma, Segal & Morrill, Inc.
New York NY
Fancy Feast

Lewis Moberly
London UK
SMA Best Friends

HARRISBERGER CREATIVE
Virginia Beach CA
Wedding Announcement
"Say I Do"

FISHER LING & BENNION
Cheltenham UK
Ward Watering Can

Desgrippes Cato Gobe & Associates

New York NY

Great American

Coleman, LiPuma, Segal & Morrill, Inc.

New York NY

Contadina

Landor Associates

San Francisco CA

Salad Bar

Morillas & Asociados
Barcelona Spain
Espuna

Harrisberger Creative
Virginia Beach VA
Fresh Salads

Ostro Design
Hartford CT
Canson Art Papers

Parham Santana Inc.
New York NY
Milliken & Co. "Airebrush"

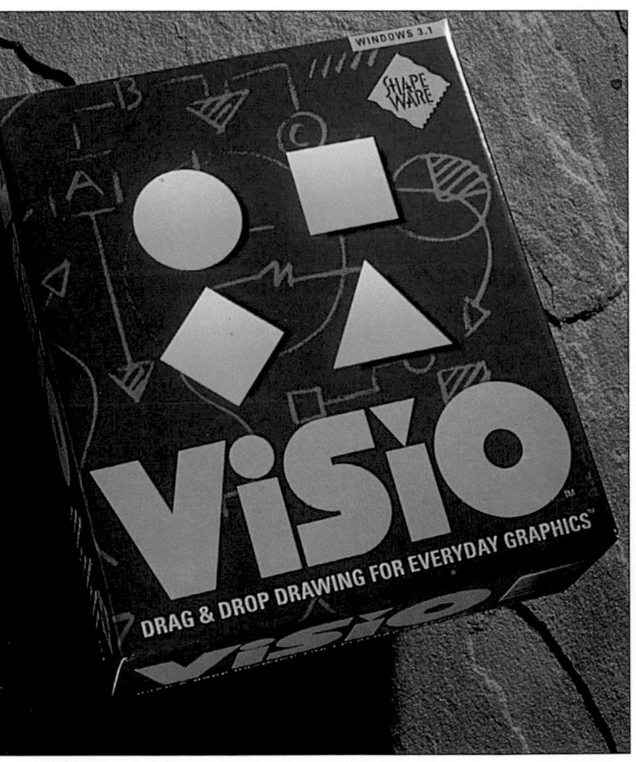

Tim Girvin Design, Inc.
Seattle WA
Visio

Sterling Design
New York NY
Walk Over

Profile Design
San Francisco CA
Approach Software

CURTIS DESIGN
San Francisco CA
Lienenkugels

THE GREAT ATLANTIC & PACIFIC TEA CO., INC.
Montvale NJ
Premium Brew

WENCEL/HESS
Chicago IL
RonRico Spiced Rum

Special

Curtis Design
San Francisco CA
Ghirardelli

Kornick Lindsay
Chicago IL
Sampoerna

Colonna Farrell Design
St. Helena CA
Carneros Alambic

Dil Consultants in Design
Sao Paulo Brazil
Brahma

Miller Sutherland
London UK
Ame

Sterling Design
New York NY
Callard & Bowser

D'Addiario Design Associates
New York NY
Paul Masson-Grape Juice

Colonna Farrell Design
St. Helena CA
Bette's Diner

Midnight Oil Studios
Boston MA
Hang Ten Sandals

B.E.P. Design Group
Brussels Belgium
Bic Fluid

Light & Coley Ltd.
London UK
Swan Vestas

J. Works

Osaka Japan

Comopolitan Chocolate

Desgrippes Cato Gobe

Paris France

Can´ Kao

Sean Michael Edwards Design, Inc.

New York NY

Bernard Genie

Gerstman & Meyers, Inc.

New York NY

NK Lawn & Garden

Profile Design

San Francisco CA

Carneros Alambic

Bailey Spiker Inc.
Norristown PA
Suede & Nubuck

Package Land Co. Ltd.
Osaka Japan
Arab Coffee

R. Bird & Company

White Plains NY

Rinso

The Weber Group, Inc.

Racine WI

Edge

Monnens-Addis Design

Emeryville CA

Tilex

Gunn Associates

Boston MA

White Rain

Elmwood

Guiseley, Leeds UK

Somerfield Furniture Polish

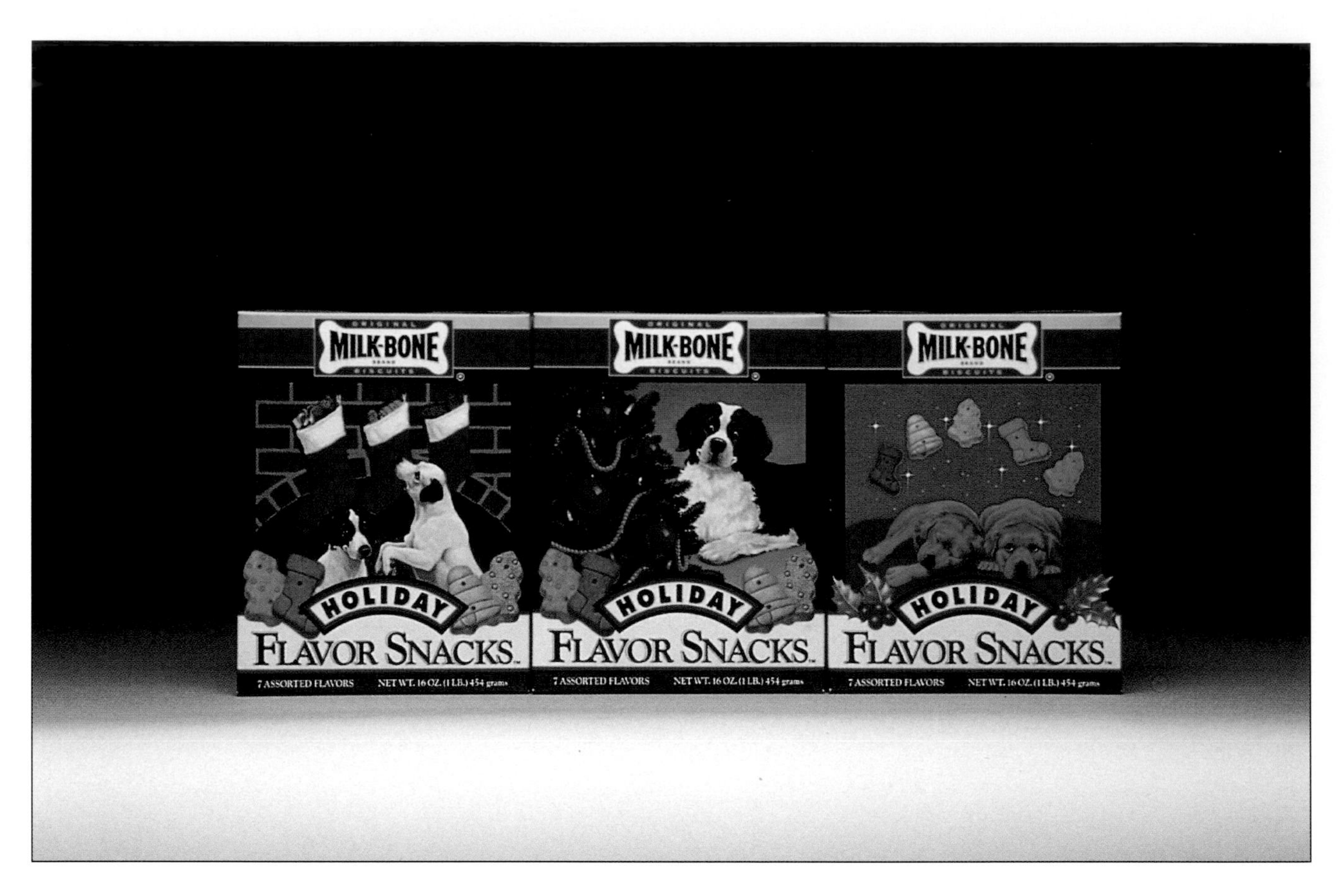

NABISCO FOODS, INC.

Parsippany NJ

Milk Bone Holiday Snacks

GERSTMAN & MEYERS, INC.

New York NY

Chico San

HILLIS MACKEY & CO., INC.

Minneapolis MN

Gobble Stix

DIL CONSULTANTS IN DESIGN
Sao Paulo Brazil
Kissy

THE SCHECHTER GROUP
New York NY
Jerky Treats

DIL CONSULTANTS IN DESIGN
Sao Paulo Brazil
Mais Mais

Gerstman & Meyers Inc.
New York NY
Sensational

The Thomas Pigeon Design Group
Toronto Canada
Hostess Munchies

Design Bridge Ltd.
London UK
Stratos

Morillas & Asociados
Barcelona Spain
Digesta

Kollberg/Johnson
New York NY
Planters Mr. Peanut

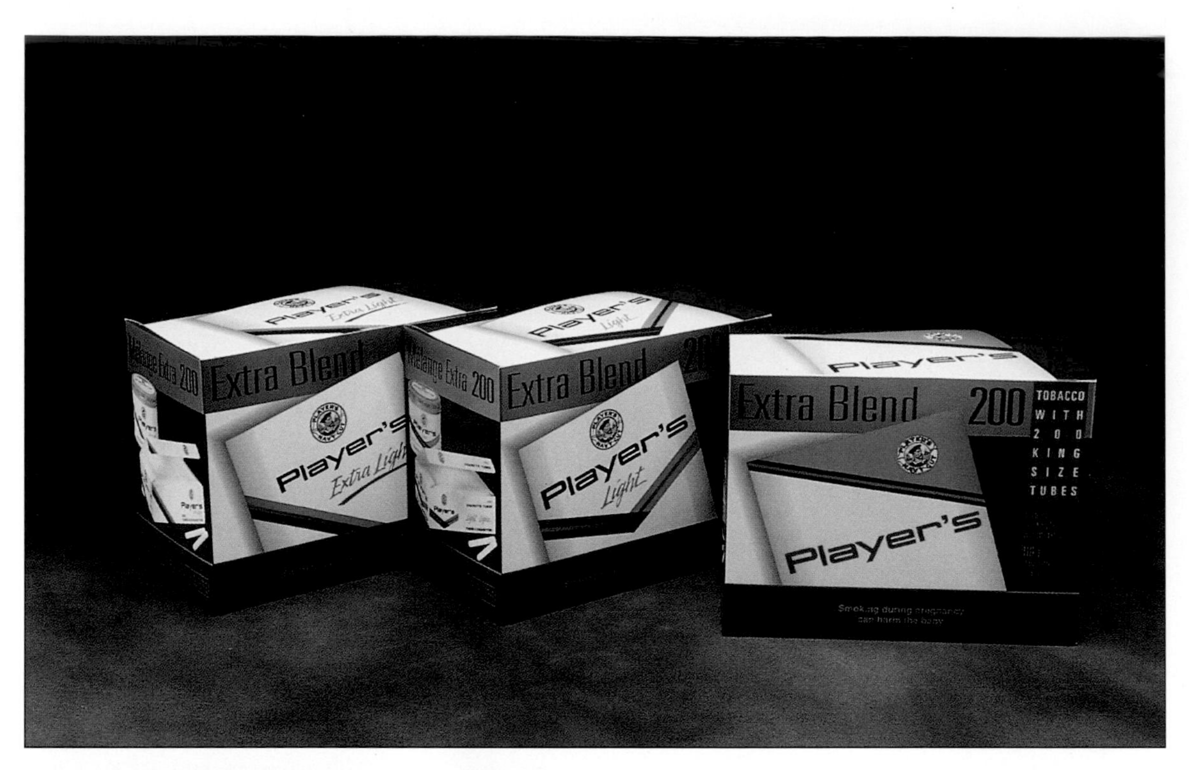

Design Partners
Toronto Canada
Player's

Marketing And
Culver City CA
Christmas Card

Parham Santana Inc.
New York NY
Jaclyn Smith Video

Hornall Anderson Design

Seattle WA

Intel Mobile Computer Peripheral Components

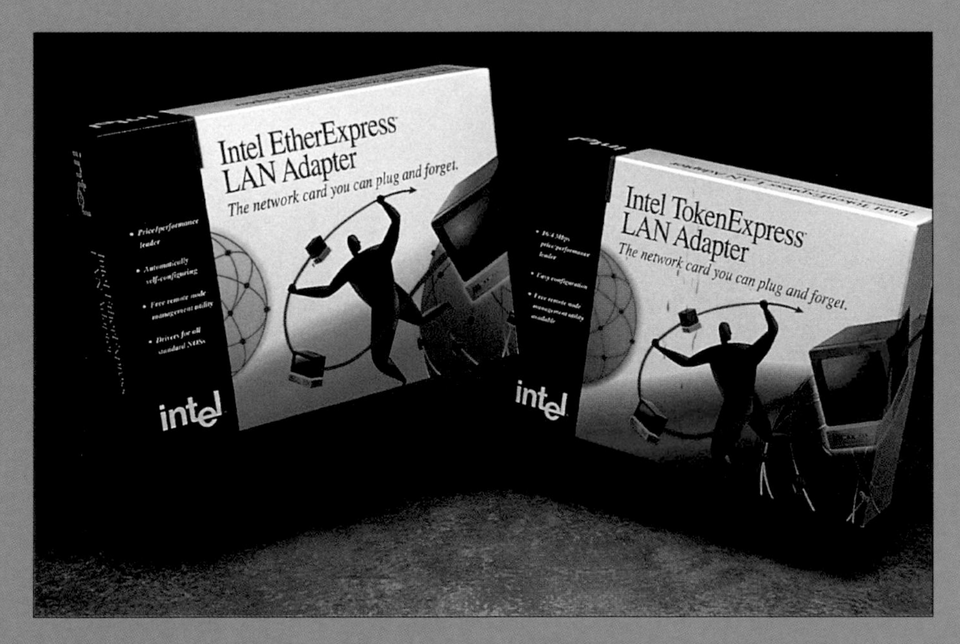

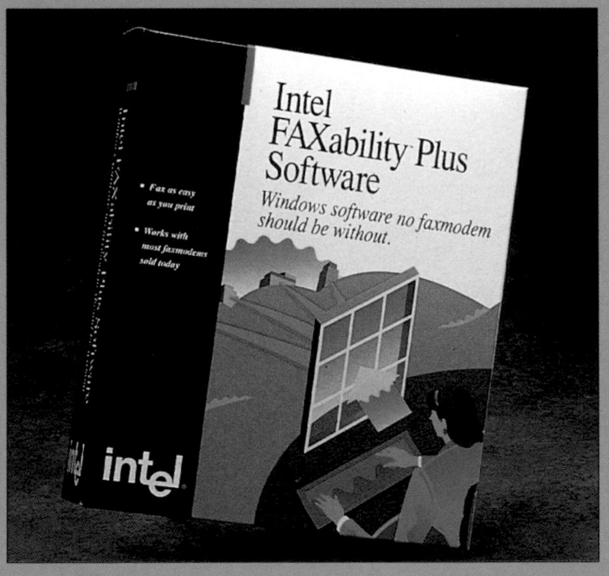

Siebert/Head Ltd.
London UK
Ty-Phoo

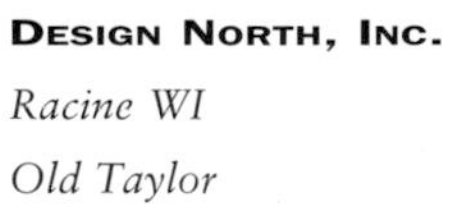

Design North, Inc.
Racine WI
Old Taylor

D'Addario Design Associates
New York NY
Crystal Ice

The Schechter Group
New York NY
Splash

WBK Design
Cincinnati, OH
$H_2Oh!$

Lipson • Alport • Glass & Associates
Northbrook IL
Hinckley & Schmitt

Desgrippes Cato Gobe
Paris France
Gervais Danone

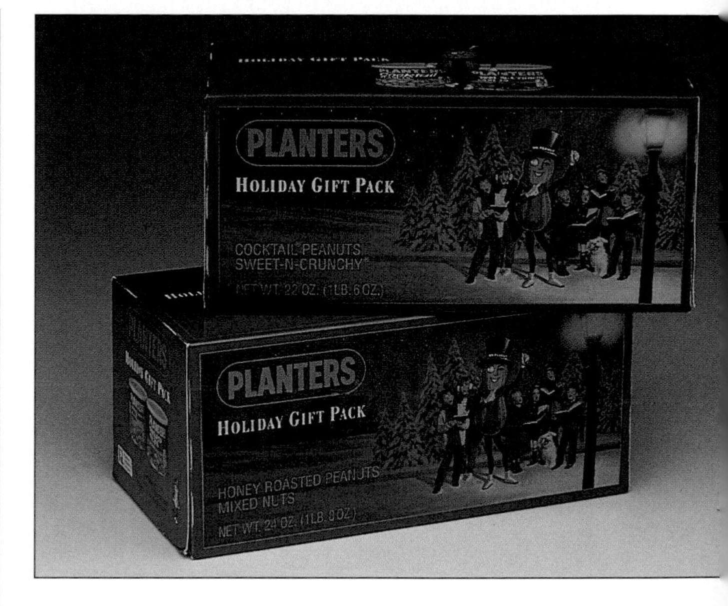

Kollberg/Johnson
New York NY
Planters' Gift Pack

Dil Consultants in Design
Sao Paulo Brazil
Sadia (Italian)

Gerstman & Meyers, Inc.
New York NY
Shenandoah

The Green House
London UK
My Little Pony

Hasbro Inc.
Pawtucket RI
Monster Face

Hasbro Inc.
Pawtucket RI
Electronic Duke

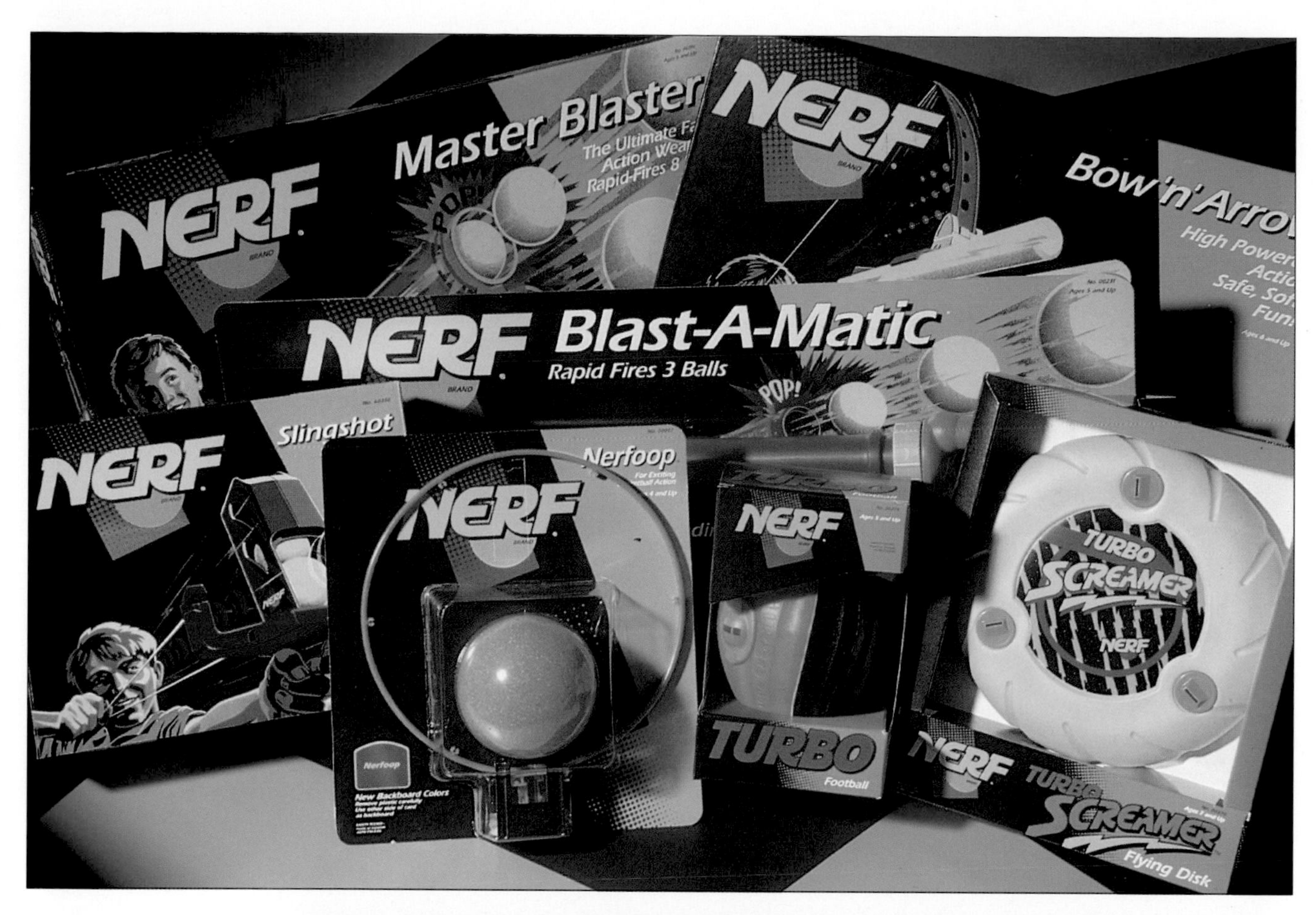

LIBBY PERSZYK KATHMAN

Cincinnati OH

Nerf

HASBRO INC.

Pawtucket RI

Cowboys

PROFILE DESIGN
San Francisco CA
Tomé

KORNICK LINDSAY
Chicago IL
Coca-Cola Classic

THE THOMAS PIGEON DESIGN GROUP
Toronto Canada
Diet Loeb

Dil Consultants in Design
Sao Paulo Brazil
Skol

Lewis Moberly
London UK
Zeiss

The Schechter Group

New York NY

Hormel Foods Agri-Nutrition

Moonink Communications

Chicago IL

Toasted Oatmeal

Hillis Mackey & Co., Inc.

Minneapolis MN

Real Fruit

B.E.P. Design Group

Brussels Belgium

Miracoli

Morillas & Asociados

Barcelona Spain

Marbu Fibra

PROFILE DESIGN

San Francisco CA

Energize

THE WEBER GROUP, INC.

Racine WI

Astro Dudes

Forward Design

Rochester NY

Kodak Writable CD

Jager Di Paola Kemp Design

Burlington VT

White Crow Office Manager

Landor Associates
San Francisco CA
Absolut Kurant

Coley Porter Bell
London UK
Babycham

The Benchmark Group
Westport CT
Godiva

Wencel/Hess

Chicago IL

Old Grand-Dad

Colonna Farrell Design

St. Helena CA

Monte Volpe

Forward Design

Rochester NY

Kodak Panther 100X

Goldsmith Tamasaki Specht Inc.

Chicago IL

Lloyd's

The Van Noy Group
Torrance CA
Kwikset Titan

Forward Design
Rochester NY
Kodak Vericolor 400

Hans Flink Design, Inc.
White Plains NY
Sunkist

Siebert/Head Ltd.
London UK
Maxwell House
Cappuccino

Elmwood
Guiseley, Leeds UK
Sangiovese

Coleman, LiPuma, Segal & Morrill, Inc.
New York NY
Hood Select

Mittleman/Robinson
New York NY
General Foods International Coffee Tins

Coleman, LiPuma, Segal & Morrill, Inc.

New York NY

Ortega

H.P. Hood Inc.

Boston MA

Hood Yogurt Combo

Morillas & Asociados

Barcelona Spain

Royal

Kraft General Foods

Glenview IL

Touch of Butter

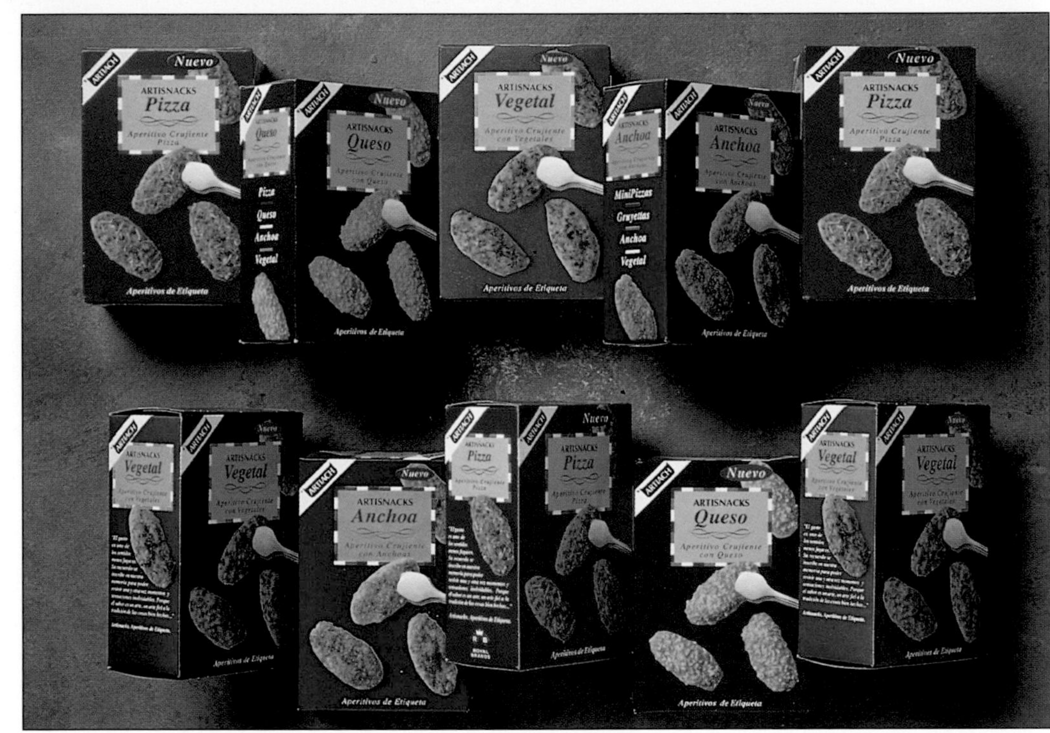

Morillas & Asociados

Barcelona Spain

Artisnacks

The Thomas Pigeon Design Group
Toronto, Ontario Canada
Primo

Wallner Harbauer Bruce
Chicago IL
Kraft Marshmallow Treat Mix

B.E.P. Design Group
Brussels Belgium
Vissticks Fishsticks

Profile Design
San Francisco CA
Lucerne

Monnens-Addis Design
Emeryville CA
Dreyer's Tropical Fruit Bars

SUNTORY LIMITED
Osaka Japan
Fruit Water

PETERSON & BLYTH ASSOCIATES
New York NY
Pepsi Crystal

The Thompson Design Group
San Francisco CA
Summit Winery

Fisher Ling & Bennion
Cheltenham UK
Finest Old

Davies Hall
London UK
Piro

Revlon, Inc.
New York NY
Revlon Outrageous

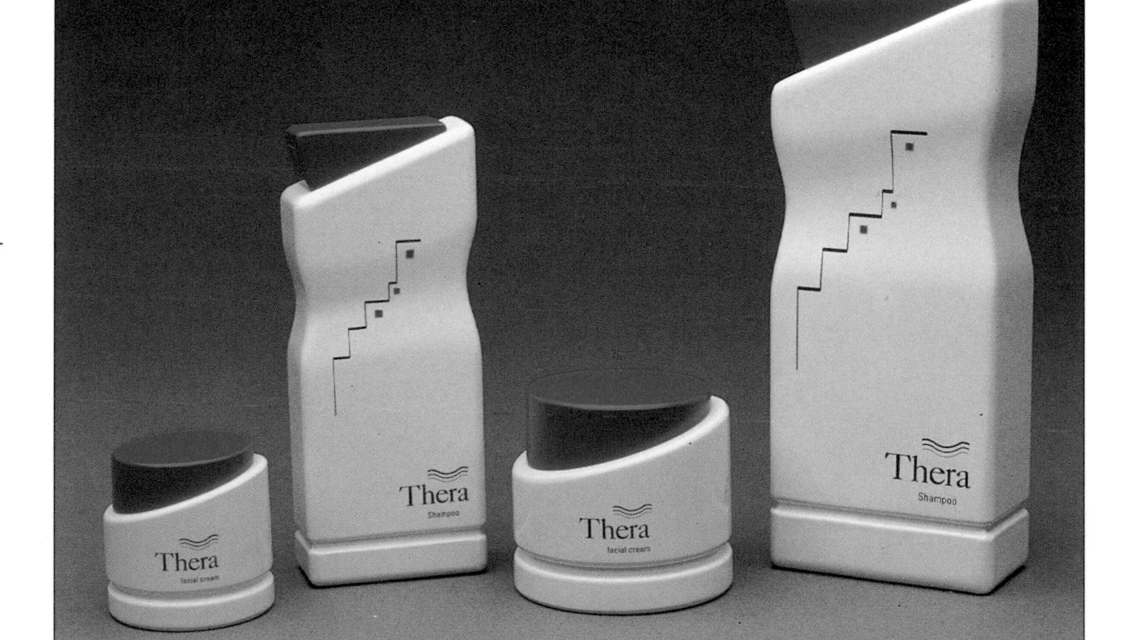

Henrik Olsen/Art Center College
Pasadena CA
Thera

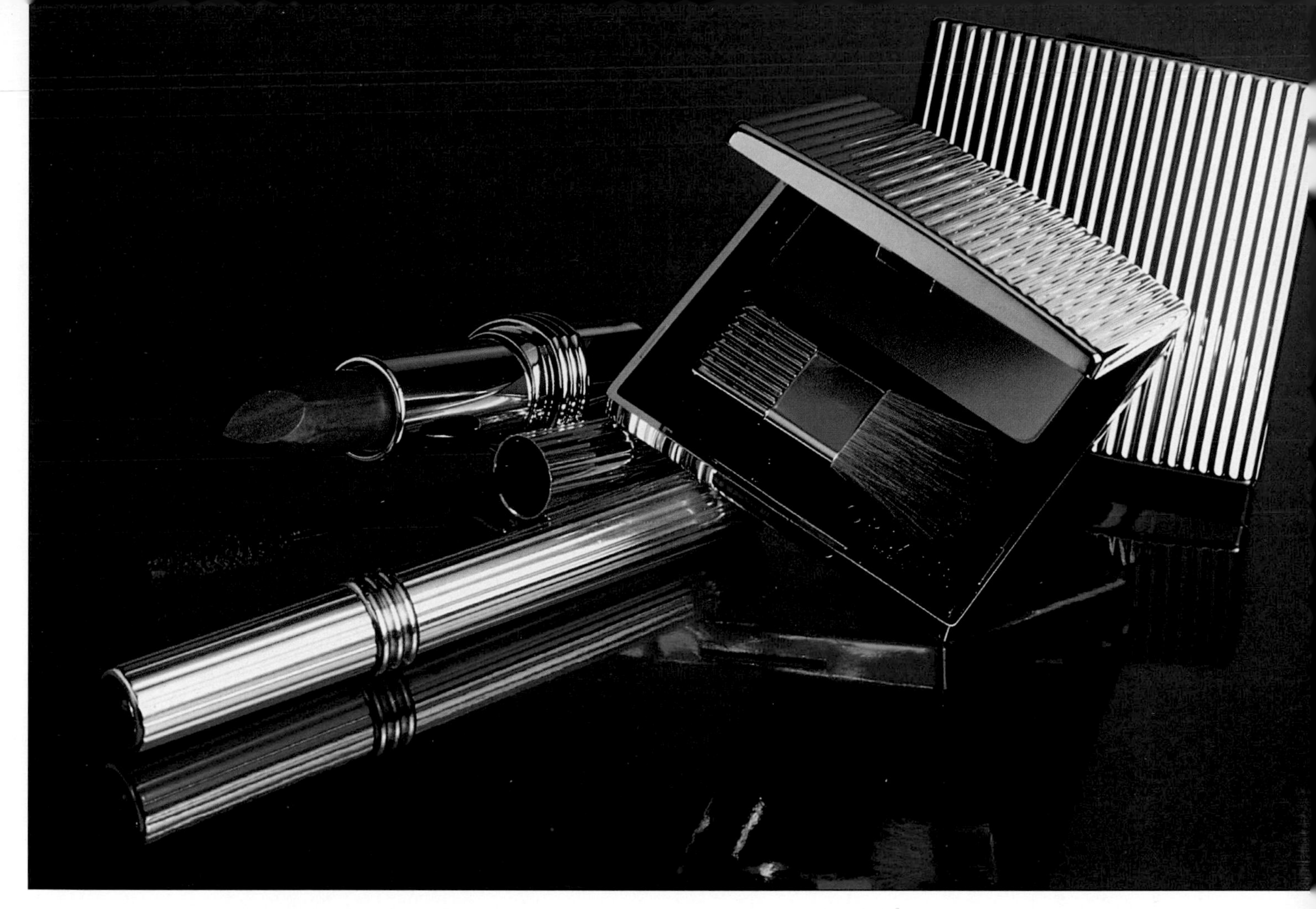

Desgrippes Cato Gobe

Paris France

Orlane

Kose Corporation

Tokyo Japan

La Valliere

Goldsmith Yamasaki Specht, Inc.

Chicago IL

3M

Source/Inc.
Chicago IL
Corporate ID Brochure

Neumeier Design Team
Palo Alto CA
Apple Font Pak

Lipson • Alport • Glass & Associates
Northbrook IL
PDC Showcase

Parham Santana Inc.
New York NY
Pantone

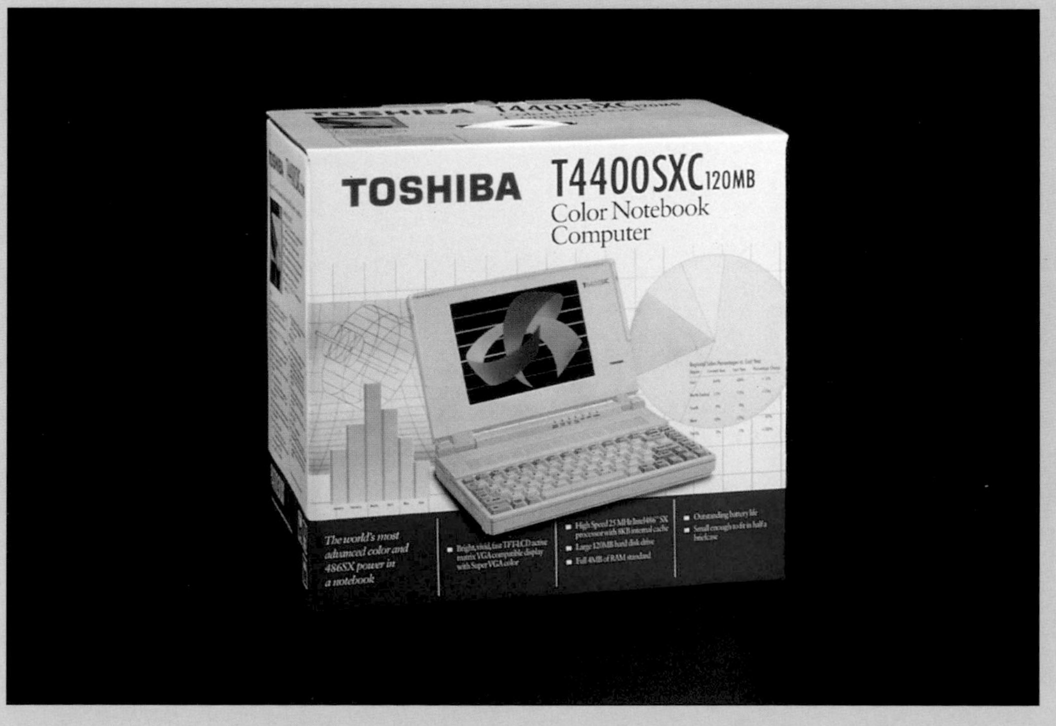

Harte Yamashita & Forest
Los Angeles CA
Toshiba

Fisher Ling & Bennion
Cheltenham UK
Yoplait

Nabisco Foods, Inc.
Parsippany NJ
Fleischmann's

PINEAPPLE DESIGN S.A.
Brussels Belgium
Bolachas De Agua

PACKAGE LAND CO., LTD.
Osaka Japan
Ka-Ni-No-Me

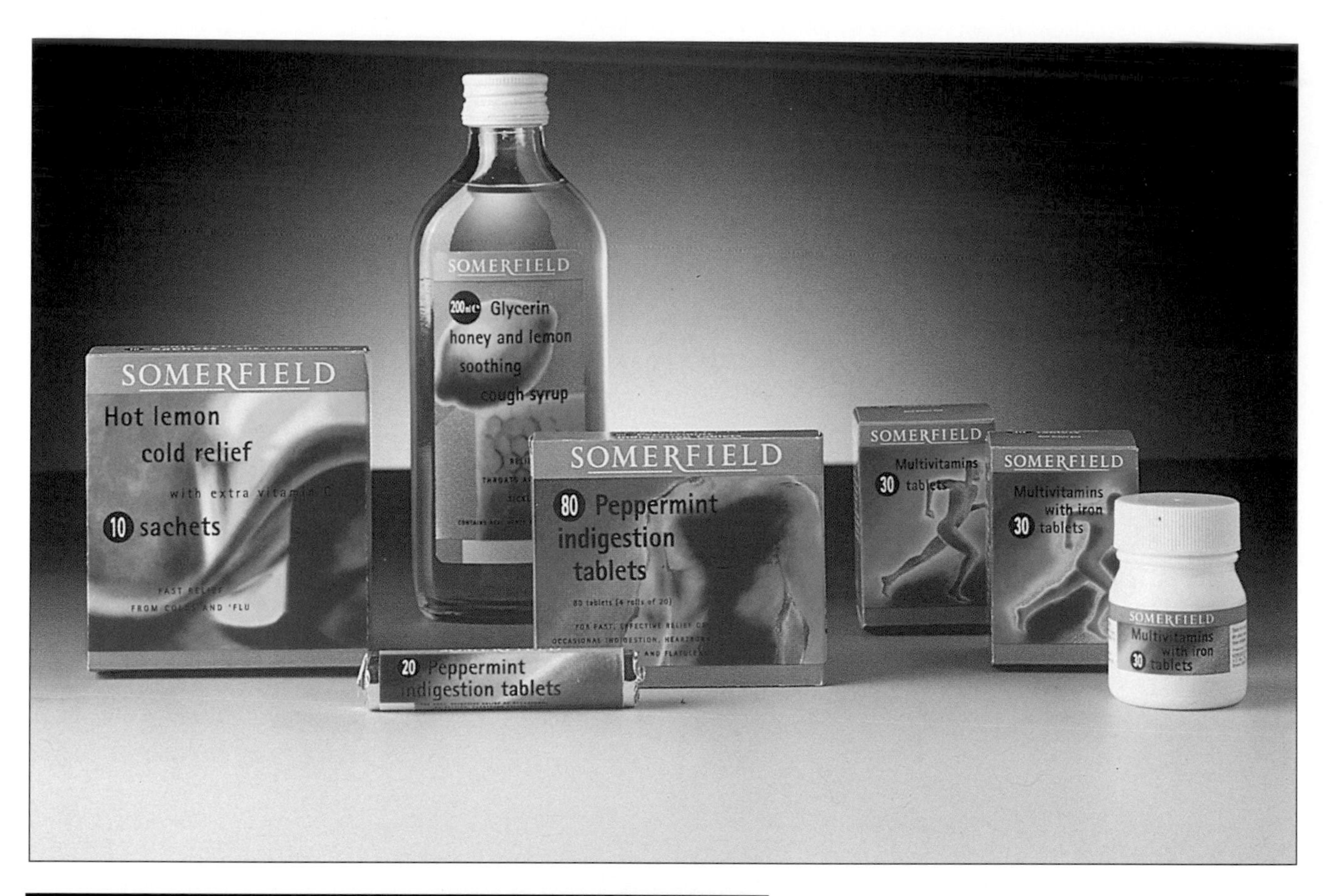

Elmwood
Guiseley, Leeds UK
Somerfield

Colonna Farrell Design
St. Helena CA
Canadian Arctic

The Benchmark Group
Westport CT
Wild Turkey

Libby Perszyk Kathman
Cincinnati OH
Seneca

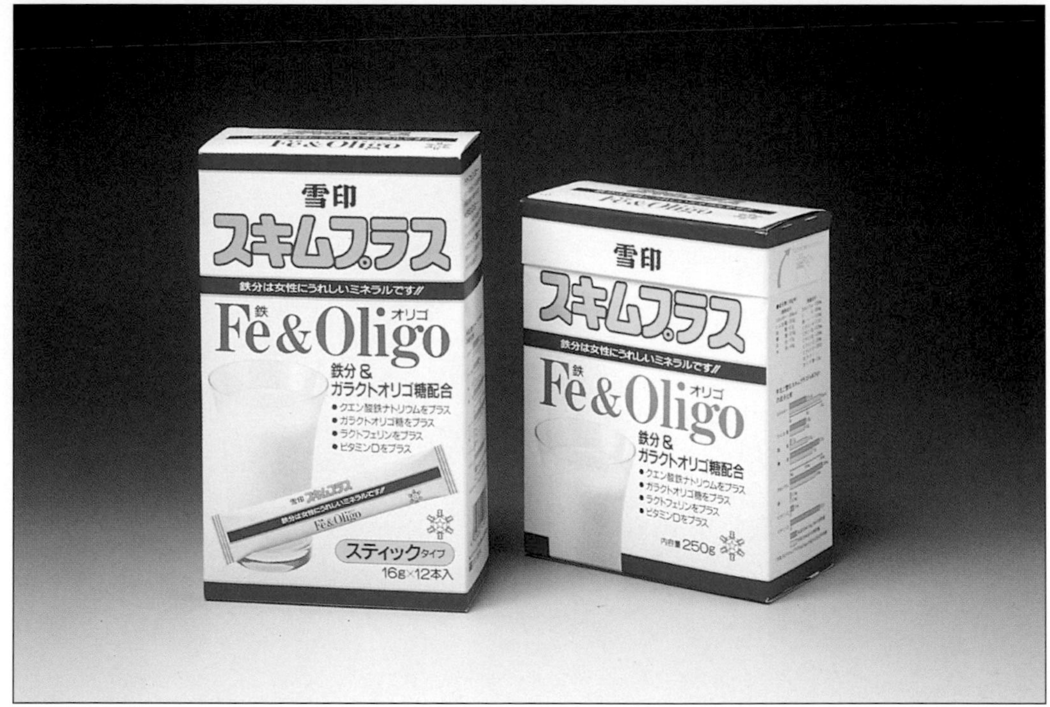

Yao Design International Inc.
Tokyo Japan
Fe & Oligo

Charles Zunda Design Co., Inc.
Greenwich CT
Binaca

Kollberg/Johnson
New York NY
Playtex

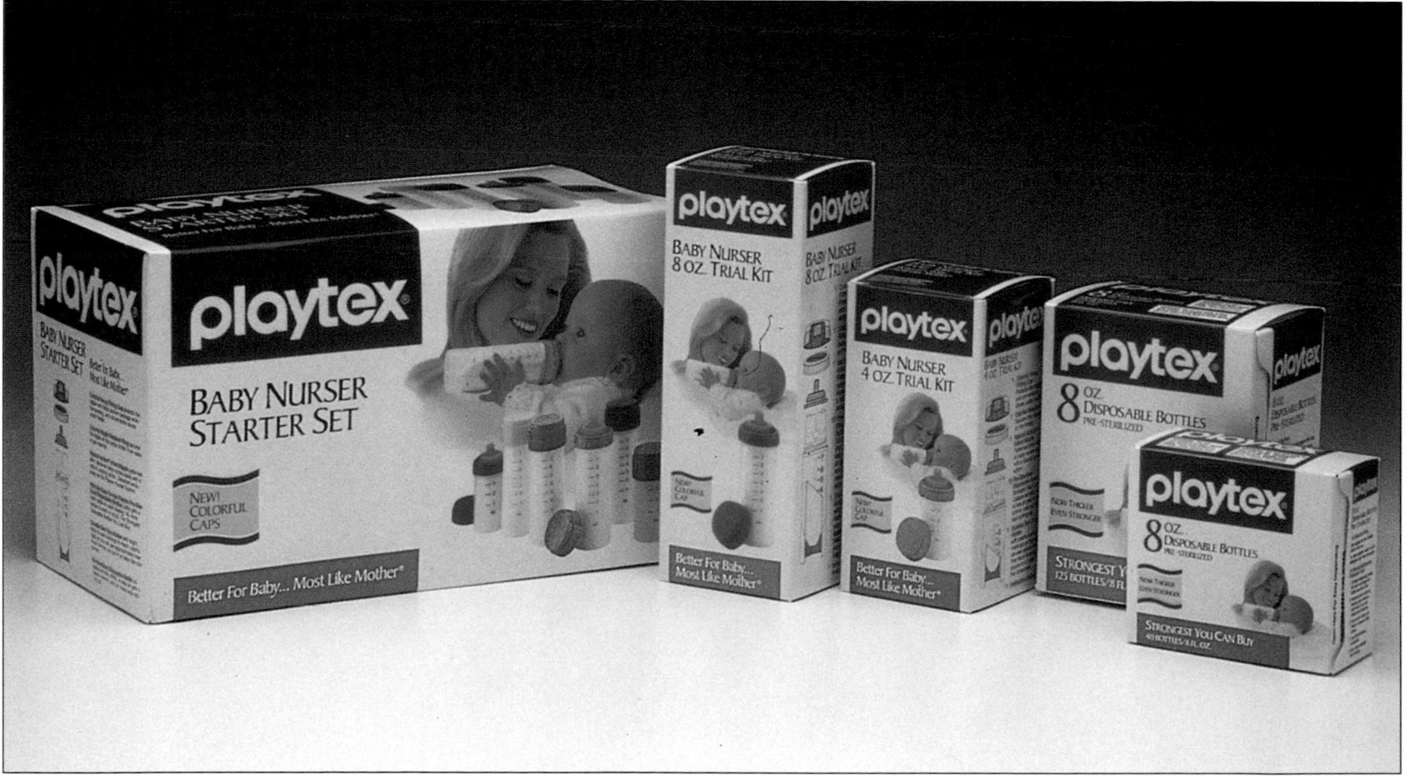

Hans Flink Design, Inc.
White Plains NY
Icy Hot

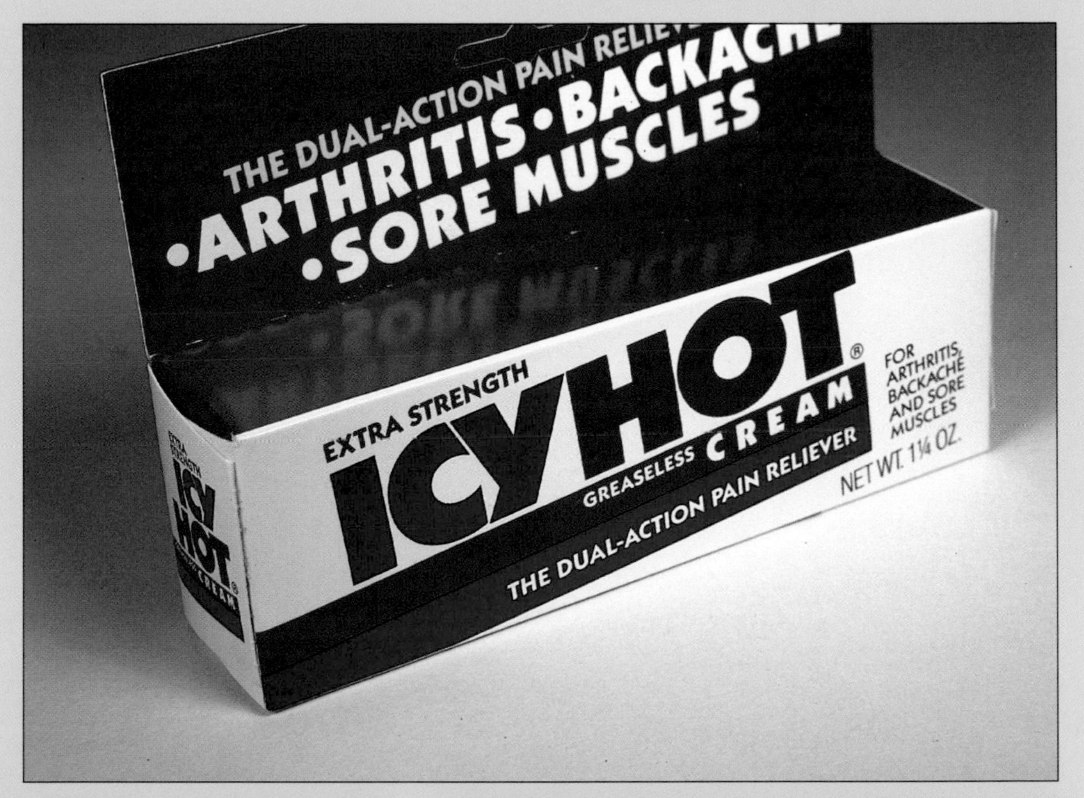

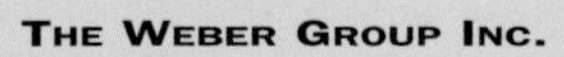

The Weber Group Inc.
Racine WI
Toilet Duck

Klaus Wuttke & Partners
London UK
Direct

Design Partners

Toronto, Ontario Canada

Maple Leaf Flakes

B.E.P. Design Group

Brussels Belgium

Soubry

Gunn Associates

Boston MA

Equal Exchange Coffee

The Schechter Group

New York NY

Hormel

TAB GRAPHICS
Denver CO
Samsonite Shadow Valet

NEUMEIER DESIGN TEAM
Palo Alto CA
Macintosh System

EARL GEE DESIGN
San Francisco CA
Perspective

Midnight Oil Studios
Boston MA
Major League Baseball Properties

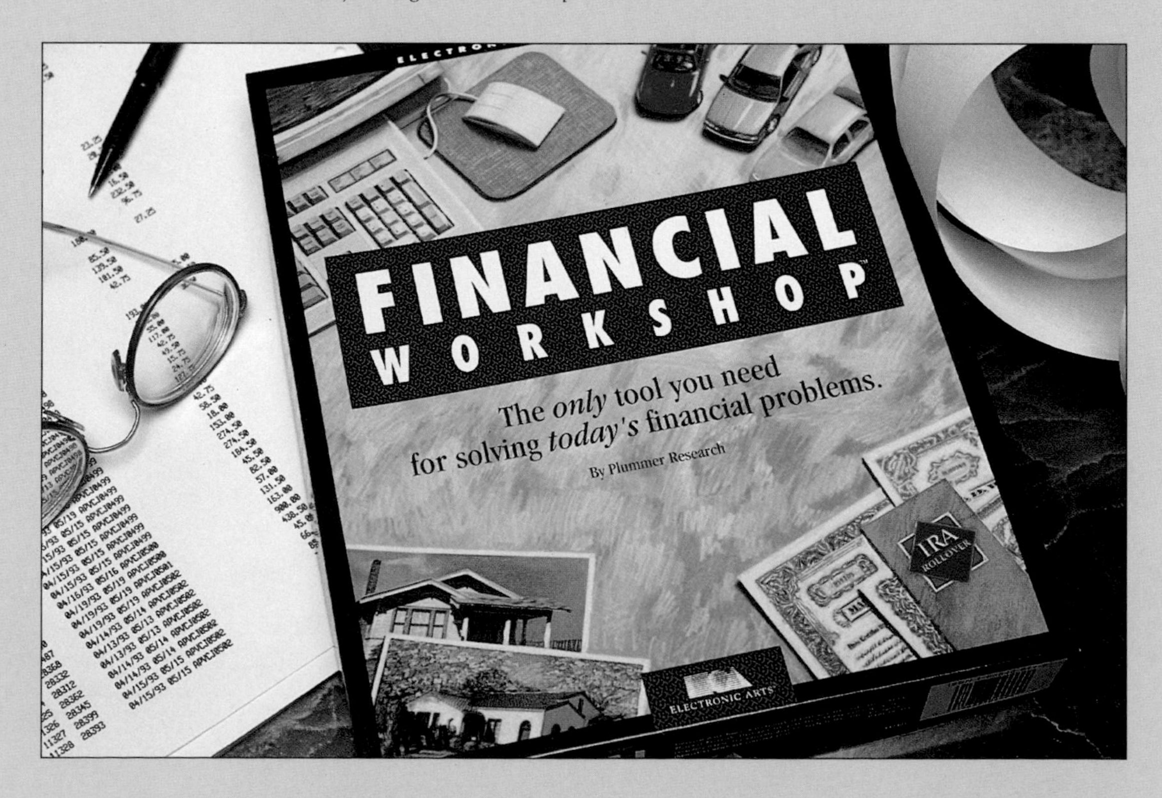

Addison Design Consultants
San Francisco CA
Financial Workshop

D'Addario Design Associates, Inc.
New York NY
Serrana

Frink Semmer & Associates, Inc.
Minneapolis MN
Land O' Lakes

ADDISON DESIGN

San Francisco CA

Rio Breeze

THE SCHECHTER GROUP

New York NY

All Sport

Landor Associates
San Francisco, CA
AR-AGE

Sean Michael Edwards Design
New York, NY
Lauder For Men

Desgrippes Cato Gobe
Paris, France
Fragonard

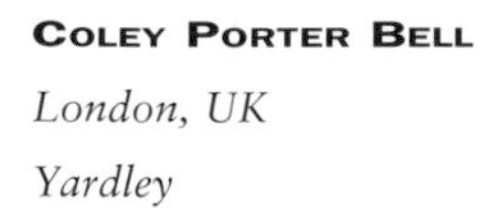

Coley Porter Bell
London, UK
Yardley

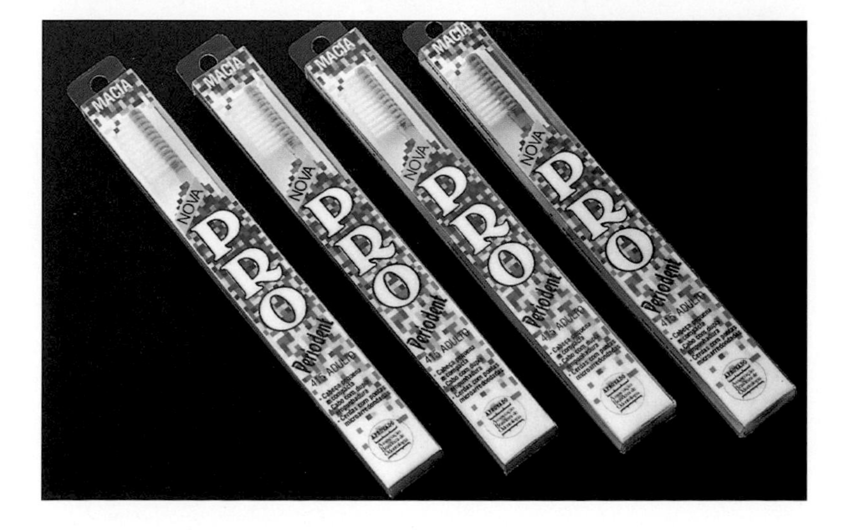

Dil Consultants in Design
San Paulo, Brazil
Pro Periodent

Design Bridge Ltd.
London UK
Jaffa Cakes

Source/Inc.
Chicago IL
Smuckers Simply Fruit

Hans Flink Design, Inc.

White Plains NY

Nestle Candy Tops

Nabisco Foods, Inc.

Parsippany NJ

Egg Beaters

Yao Design International Inc.

Tokyo Japan

Japanese Potato Chips

Coley Porter Bell

London, UK

Lego Model Team

Design North, Inc.

Racine WI

Protector Fly Trap

Earl Gee Design

San Francisco CA

Claris Clear Choice

Neumeier Design Team

Palo Alto CA

Kodak Shoebox

Addison Design

San Francisco CA

Disc Drive Kit

Kornick Lindsay
Chicago IL
American Orchards
Heublein

Special

Davies Hall
London UK
Biere De Garde

Wencel/Hess
Chicago IL
Kessler

Fisher Ling & Bennion
Cheltenham UK
Woodward's Herbal Baby Drink

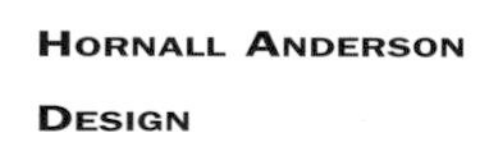

Hornall Anderson Design
Seattle WA
Starbucks Coffee

ABKCO Films
1700 Broadway
New York, NY 10019

Addison Design
575 Sutter Street
San Francisco, CA
94102

American Packaging Corp.
125 W. Broad Street
Story City, IA 50248

Primo Angeli, Inc.
590 Folsom Street
San Francisco, CA
94105

Apple DesignSource
311 East 46th Street
New York, NY 10017

Barrett Design
545 Concord Avenue
Cambridge, MA 02138

The Benchmark Group
265 Post Road West
Westport, CT 06880

B.E.P. Design Group
Rue Des Mimosas 44
B-1030 Brussels,
Belgium

Besser Joseph Partners
1546 7th Street
Santa Monica, CA
90401

Biegler Design
1030 Collage Avenue
Wheaton, IL 60187

R. Bird & Company
11 Martine Avenue
White Plains, NY
10606

Howard Blonder & Associates
10933 Lakewood Boulevard
Downey, CA 90241

The Boardroom Design Group
25825 Science Park Dr.
Cleveland, OH 44122

Bright & Associates
901 Abbot Kinney Boulevard
Venice, CA 90291

Britton Design
164 W. Spain Street
Sonoma, CA 95476

Wallace Church Associates
330 East 48th Street
New York, NY 10017

Clark/Linsky Design
201 Main Street
Charlestown, MA
02129

Coleman, LiPuma, Segal & Morrill
305 East 46th Street
New York, NY 10017

Coley Porter Bell
4 Flitcroft Street
London WC2H 8DJ
U.K.

Colgate Palmolive Company
300 Park Avenue
New York, NY 10022

Colonna Farrell Design
1335 Main Street
St. Helena, CA 94574

Curtis Design
One Charlton Court
San Francisco, CA
94123

Helene Curtis, Inc.
325 N. Wells Street
Chicago, IL 60610

D'Addario Design Associates
123 West 44th Street
New York, NY 10036

Davies Hall
The Forum
74-80 Camden Street
London NW1 OEG
U.K.

Desgrippes Cato Gobe
18 Bis, Avenue de la Motte
Picquet Paris 75007
France

Desgrippes Cato Gobe & Associates
411 Lafayette Street
New York, NY 10003

Design Board/Behaeghel
50, av. Georges Lecointe
1180 Brussels, Belgium

Design Bridge Ltd.
18 Clerkenwell Close
London EC1R OAA
U.K.

Design Centre of Cincinnati
225 East 6th Street
Cincinnati, OH 45202

Design Forum
3484 Far Hills Avenue
Dayton, OH 45429

Design Group Italia
Vicola S. Maria alla
Porta 1
20123 Milano, Italy

Design North Inc.
8007 Douglas Avenue
Racine, WI 53402

Design One
1500 Sansome Street
San Francisco, CA
94111

Design Partners
282 Richmond Street East
Toronto, ON M5A 1P4,
Canada

Deskey Associates
145 East 32nd Street
New York, NY 10016

DiDonato Associates Inc.
811 W. Evergreen Street
Chicago, IL 60622

Dil Consultants in Design
R. Oscar Freire
379 16 and CJ.162
Cerqueira Cesar
Sao Paulo, Brazil
01426-001

Charon Dyer Design
481 Manor Ridge Drive, NW
Atlanta, GA 30305

Sean Michael Edwards
28 West 25th Street
New York, NY 10010

Elmwood
Elmwood House,
Ghyllroyd
Guiseley, Leeds L520
9BU U.K.

Fatta Design Group
100 Putnam Green
Greenwich, CT 06830

Fisher Ling & Bennion
Glenmore Lodge,
Wellington Square
Cheltenham GL50 4JY
U.K.

Hans Flink Design
11 Martine Avenue
White Plains, NY 10606

Forward Design
1115 E. Main Street
Rochester, NY 14609

Frankfurt Balkind Partners
244 East 58th Street
New York, NY 10022

Earl Gee Design
501 Second Street
San Francisco, CA
94107

Gerstman + Meyers
111 West 57th Street
New York, NY 10019

Tim Girvin Design
1601 Second Avenue
Seattle, WA 98101

Goldsmith Yamasaki Specht Inc.
900 N. Franklin Street
Chicago, IL 60610

Georges Gotlib, Inc.
515 Madison Avenue
New York, NY 10022

Graphic Edge
5981 Engineer
Huntington Beach, CA
92649

Graphique Design
555 West 57th Street
New York, NY 10019

Graphique Design
Magna House
76-80 Church Street
Staines, Middlesex
England
TW184XR UK

The Great Atlantic & Pacific Tea Co., Inc.
2 Paragon Drive
Montvale, NJ 07645

The Green House
64 High Street
Harrow-on-the-Hill
London HA1 3LL U.K.

GRP Records
555 West 57th Street
New York, NY 10019

Gunn Associates
275 Newbury Street
Boston, MA 02116

Handler Design
17 Ralph Avenue
White Plains, NY
10606

Hanson Associates, Inc.
133 Grape Street
Philadelphia, PA 19127

Harrisberger Creative
420 Investors Place
Virginia Beach, VA
23452

Harte Yamashita & Forest
5735 Melrose Avenue
Los Angeles, CA 90038

Hartmann & Mehler Designers GmbH
Corneliusstrasse 8
60325 Frankfurt am Main 1
Germany

Hasbro Inc.
1027 Newport Avenue
Pawtucket, RI 02862

Hermsen Design Associates
5151 Beltline Road
Dallas, TX 75240

Hewlett-Packard
11311 Chinden Boulevard
Boise, ID 83714

Hewlett-Packard Co.
16399 W. Bernado Dr.
San Diego, CA 92127

Hood Design Group
54 Harrison Street
Brookline, MA 02146

Hornall Anderson Design
1008 Western Avenue
Seattle, WA 98104

H.P. Hood Inc.
500 Rutherford Avenue
Boston, MA 02129

Hughes Design, Inc.
One Bishop Street
Norwalk, CT 06851

Hunt Weber Clark Design
51 Federal Street
San Francisco, CA
94107

ID exposure, Inc.
2300 E. Douglas Street
Wichita, KS 67214

Ilium Associates, Inc.
500 108th Avenue NE, #2450
Bellevue, WA 98004

Jager Di Paola Kemp Design
308 Pine Street
Burlington, VT 05401

Baker Jazdzewski
6 Dean Street
London W1V 5RN U.K.

Libby Perszyk Kathman
19 Garfield Place
Cincinnati, OH 45202

Mary Kay Cosmetics, Inc.
8787 Stemmons Freeway
Dallas, TX 75247

Ketchum Promotions
16 West 22nd Street
New York, NY 10010

Kollberg/Johnson
7 West 18th Street
New York, NY 10011

Kornick Lindsay
161 E. Erie Street
Chicago, IL 60611

Kose Corporation
Packaging Design Section
1-9-9 Hatchobori
Chuo-Ku, Tokyo 104
Japan

Kraft General Foods
One Kraft Court
Glenview, IL 60025

Landor Associates
1001 Front Street
San Francisco, CA
94111

Nick Lane Design
46 Varda Landing
Sausalito, CA 94965

The Leonhardt Group
1218 3rd Avenue,
Suite 620
Seattle, WA 98040

Light & Coley Ltd.
20 Fulham Broadway
London SW6 1AH U.K.

Lindt & Sprungli
One Fine Chocolate Place
Stratham, NH 03885

Lipson•Alport•Glass & Associates
666 Dundee Road, 103
Northbrook, IL 60062

Lister Butler Inc.
475 Fifth Avenue
New York, NY 10017

Hillis Mackey & Co.
1550 Utica Avenue South
Minneapolis, MN
55416

Marketing And
8522 National Boulevard
Culver City, CA 90232

Marketing, Visuals & Promotions
111 Third Avenue South
Minneapolis, MN
55401

MCI Design GmbH
Corneliusstrabe 8
6000 Frankfurt am Main 1
Germany

Midnight Oil Studios
51 Melcher Street
Boston, MA 02210

Millen & Ranson
285 Broadway
New York, NY 10013

Mittleman/Robinson
3 West 18th Street
New York, NY 10011

Lewis Moberly
33 Gresse Street
London W1P 1PN U.K.

Monnens•Addis Design
250 Emery Bay Marketplace
Emeryville, CA 94608

Moonink Communications
205 N. Michigan Avenue
Chicago, IL 60601

Morillas & Asociados
PO Santa Eulalia, 17-19
Barcelona, Spain 08017

Murrie, Lienhart, Rysner & Associates
58 West Huron Street
Chicago, IL 60610

Nabisco Foods, Inc.
7 Sylvan Way
Parsippany, NJ 07054

Neumeier Design
915 Waverly Street
Palo Alto, CA 94301

Mark Oliver, Inc.
One West Victoria Street
Santa Barbara, CA
93101

Mr. Henrik Olson
Art Center College
987 E. DelMar St., #14
Pasadena, CA 91106

Ostro Design
147 Fern Street
Hartford, CT 06105

Package Design of
America
955 Connecticut Avenue
Bridgeport, CT 06607

Package Land Co., Ltd.
201 Tezukayama Tower
Plaza
1-3-2 Tezukayama-
NAKA
Sumiyoshi-ku, Osaka
558 Japan

Peterson & Blyth
216 East 45th Street
New York, NY 10016

Pethick & Money
Studios 4 & 5
75 Filmer Road
London SW6 7JF U.K.

The Thomas Pigeon
Design Group
234 King Street East
Toronto, ON M5A 1K1,
Canada

Pineapple Design S.A.
Rue de la
Consolation 56
1030 Brussels, Belgium

William Plewes Design
102 Atlantic Avenue
Toronto ON M6K 1X9
Canada

Prepco
6190 E. Slauson Avenue
Los Angeles, CA 90040

Profile Design/KDF
Planet
151 Townsend Street
San Francisco, CA
94107

Mike Quon Design
Office
568 Broadway
New York, NY 10012

Ratta Design
Communications
22 Free Street
Portland, ME 04101

Reed & Barton
144 W. Britannia Street
Taunton, MA 02780

Revlon, Inc.
625 Madison Avenue
New York, NY 10022

Roy Ritola Inc.
431 Jackson Street
San Francisco, CA 94111

Parham Santana
7 West 18th Street
New York, NY 10011

David Scarlett &
Associates
401 Lafayette Street
New York, NY 10003

Schafer Associates, Inc.
635 Butterfield Road
Oak Brook Terrace, IL
60181

The Schechter Group
212 East 49th Street
New York, NY 10017

Peter Seitz & Associates
4633 Sunset Ridge
Golden Valley, MN
55416

Frink Semmer and
Associates
505 East Grant Street
Minneapolis, MN
55404

Robin Shepherd Studios
476 Riverside Avenue
Jacksonville, FL 32202

Siebert/Head Limited
38 Hans Crescent
Knightsbridge, London
SW1X OLZ U.K.

Silver Burdett & Ginn
160 Gould Street
Needham, MA 02194

Karen Skunta &
Company
1382 West 9th Street
Cleveland, OH 44113

Agnew Moyer Smith
503 Martindale Street
Pittsburgh, PA 15212

Source/Inc.
116 S. Michigan Avenue
Chicago, IL 60603

Bailey Spiker Inc.
805 E. Germantown
Pike
Norristown, PA 19401

Michael Stanard, Inc.
1000 Main Street
Evanston, IL 60202

Sterling Design
800 Third Avenue
New York, NY 10022

Suntory Limited
40, Dojimahama 2-1
Kitaku, Osaka, Japan
530

Supon Design Group
1000 Connecticut
Avenue NW, #415
Washington, DC 20036

Miller Sutherland
6 D'Arblay Street
London W1V 3FD U.K.

TAB Graphics
1120 Lincoln Street,
#700
Denver, CO 80203

Tactix Communications
& Design
158 Coach Hill Drive
Kitchener ON N2E 1P4
Canada

TDC/The Design
Company
165 Page Street
San Francisco, CA
94102

The Thompson Design
Group
524 Union Street
San Francisco, CA
94133

Tulocay Design
Partnership
1475 Fourth Street
Napa, CA 94559

United States Surgical
Corporation
150 Glover Avenue
Norwalk, CT 06856

The Van Noy Group
19750 S. Vermont
Avenue
Torrance, CA 90502

Verge Lebel
Communication Inc.
1039, Av. Deserabies
Quebec, Canada
G1R ZN1

Wallner Harbauer Bruce
500 N. Michigan Ave.
Chicago, IL 60611

Walsh & Associates
4464 Fremont Avenue
North
Seattle, WA 98103

WBK Design
19 Garfield Place
Cincinnati, OH 45202

The Weber Group, Inc.
3001 Washington Ave.
Racine, WI 53405

Wencel/Hess
320 N. Michigan Avenue
Chicago, IL 60601

Werbin Associates
895 Port Drive
Mamaroneck, NY
10543

White Tiger Inc.
709 White Horse Pike
Audubon, NJ 08106

J. Works Co., Ltd.
Nishikawa Bldg. 6F
2-3-22 Tenjinbashi
Kita-Ku
Osaka 530 Japan

Klaus Wuttke &
Partners
5-6 Clipstone Street
London W1P 7EB U.K.

Yao Design International
Shoheikan Bldg. 7F
4 Honshiocho,
Shinjuku-ku, Tokyo 160
Japan

Yonetsu Design
2-4-16 Shimizu
Suginami-Ku
Tokyo 167 Japan

Charles Zunda Design
80 Mason Street
Greenwich, CT 06830